28·69

Oil Pollution and the PublicInterest:

A STUDY OF THE SANTA BARBARA OIL SPILL

A. E. KEIR NASH

Department of Political Science
University of California, Santa Barbara

DEAN E. MANN

Department of Political Science
University of California, Santa Barbara

PHIL G. OLSEN

Department of Geology
Santa Barbara City College

INSTITUTE OF GOVERNMENTAL STUDIES

University of California, Berkeley

1972

LIBRARY OF CONGRESS CATALOGING IN PUBLICATION DATA

Nash, A.   E.   Keir
    Oil pollution and the public interest.

    1. Marine pollution--Santa Barbara Channel.
2. Oil pollution of rivers, harbors, etc.--Santa
Barbara Channel. I. Mann, Dean E., joint author.
II. Olsen, Phil G., joint author. III. Title.
TD427.P4N37   301.31   72-5116
ISBN 0-87772-085-1

$3.75

Contents

Foreword

Can man adequately control his ever-more-powerful technology? Can he intelligently guide both its impact on the environment and its ever-increasing demands on available energy resources? Except for learning to govern himself in relationship with his fellow human beings, man's way of dealing with the environmental and energy crises may be his most important task in the closing decades of this century.

The Santa Barbara oil spill poses many of the issues in microcosm, and the work of A.E. Keir Nash, Dean E. Mann and Phil G. Olsen presents a careful, thoughtful and imaginative analysis of recent developments. While events related to the Santa Barbara disaster are by no means conclusive, they highlight the stress, struggle, and trauma that accompany attempts to adapt an inherited and primarily exploitative system of resource management to the pressing needs of a new era.

The difficulties of the adaptive effort were compounded by the complexity of the Santa Barbara case and of the regulatory machinery that governs coastal petroleum. Virtually every arm and level of government was involved. The county acted as the agent of those who would save Santa Barbara from future disasters. The United States Secretary of the Interior, of course, had primary responsibility for administering and encouraging the process of exploitation, as well as for regulating it. The State of California controlled important tidelands, which are administered by the State Lands

Commission. Some influences came from farther afield,
such as the international system of petroleum pricing
and quota-setting that determined many of the priorities
for development of underwater petroleum reserves like
those lying beneath Santa Barbara Channel. Additional
domestic institutional involvements were those of the
courts, both houses of Congress, the Office of the President, and the California State Executive.

This examination of the Santa Barbara oil spill,
and of what the various actors were trying to do, affords
an instructive case study of institutional response in a
complex new situation. The episode is significant.
With energy demands burgeoning faster than either the
population or the economy--at the same time that pressures for environmental protection and preservation are
also rising--the future seems likely to bring a clash
of major proportions.

Events like the Santa Barbara spill may be preliminary skirmishes in the coming battles. If we can command sufficient vision and understanding to fathom the
significance of these and similar developments, perhaps
we can work out imaginative new solutions, and avoid
far more damaging environmental and energy conflicts.
But some changes will have to be made in how we deal
with these crucial policy issues, and future institutional responses will need to be far more pliant and
flexible, relevant and effective than much of the organ-
izational creaking and groaning that accompanied the
Santa Barbara oil spill.

The authors concluded their final chapter by making
a guarded prognostication for the longer run. It seems
appropriate to conclude this Foreword with the following
excerpt:

> To sum up, it is too early to conclude
> that Santa Barbarans will in this case
> prevail over the awesome power of the
> oil industry. Yet it is also too

early to conclude that in the foreseeable future the balance between
competing values and interests will
remain where they have characteristically been since the New Deal.
The importance of the Santa Barbara
oil spill and its local aftermath
lies not in its representing a
last turning-point for reevaluating
priorities, but rather in its clear
illumination in the public limelight of the emerging political
competition between old and new
economic and ecological life-styles.
The Santa Barbara incident is
weighted with significance for the
future functioning of the American
political system.

Nancy Kreinberg was responsible for the careful
editing of the manuscript, and Judy Rasmussen did the
typing.

Stanley Scott
Assistant Director

Acknowledgments

In the course of writing this monograph, the authors have accumulated a list of debts to numerous individuals who have lent their aid and given their time in sundry ways. We are especially grateful for the help of the following persons: Joseph Foster, research assistant; Senator Robert J. Lagomarsino, of the California Legislature; Eleanor L. Leary, office of United States Senator Stuart Symington; Theodore M. Leary, Jr., legislative assistant to United States Senator Abraham Ribicoff; Marvin Levine, Deputy Santa Barbara County Counsel; California State Assemblyman W. Don MacGillivray; David Minier, Santa Barbara District Attorney; United States Senator Robert Packwood; Paul Rosten, petroleum engineer, Santa Barbara; Fred Schambeck, Deputy Regional Supervisor, Western Region, United States Geological Survey; Robert Sharp, petroleum engineer, Santa Barbara; Lois Sidenberg, "Get Oil Out;" Robert E. Sollen, reporter, Santa Barbara *News-Press*; David Stanton, Deputy Attorney General, Los Angeles Office, California Department of Justice; and United States Congressman Charles Teague.

CHAPTER I

Introduction

THE SPILL

9:00 A.M., January 28, 1969: the masts of over 700 yachts tilted almost imperceptibly back and forth in the gentle winter swell of Santa Barbara's palm-fringed harbor. Five miles out to sea, an oil-drilling crew had just reached the maximum projected depth of 3,479 feet for what promised to be the fifth productive development well on Offshore Platform A, belonging to the Union Oil Company of California and its three partners: Mobil, Texaco and Gulf. Eighty miles away in Union's executive offices 12 stories above Los Angeles' freeway fumes, company president Fred Hartley might at that moment have reflected pleasurably, anticipating imminent returns from almost nine square miles of Pacific Ocean floor. This was Parcel 402 to which Union and her three partners[1] had acquired drilling rights for $61,418,000 not twelve months before.

On February 6, 1968, at the Los Angeles Biltmore the federal government had, with abundant satisfaction, disclosed the total sum that 12 major oil companies and several minor companies had bid for leasing 110 parcels of the Outer Continental Shelf in the Santa Barbara Channel. The successful high bidders had offered a record bonus of $602,719,261.60 plus $3 per annum per acre and 16.67 percent of future production royalties. Such a sum indicated considerable industry confidence that substantially more than met the scuba diver's eye lay beneath the Channel sands. Yet, in the intervening months most of the lessees seemed to have met with something less than quick returns: By early 1969 the oil companies had invested almost $1 billion but had announced successes on only two leases.

By contrast, Parcel 402, lying quietly 45 fathoms below the deck of Platform A, promised to be richly rewarding. From this parcel alone Union had a good chance of recovering most of her total investment of $76.5 million[2] in, to use the industry's vernacular, Santa Barbara's whale pasture. Matters had proceeded smoothly enough for Donald W. Solanas, Western District Regional Oil and Gas Supervisor of the U.S. Geological Survey, to grant Union a variance from the Department of the Interior's standard offshore casing requirements.

On wells as deep as A-21, federal regulations called for a standard minimum of 300 feet of *conductor casing*--the first string of protective casing normally set beneath the ocean floor--and a blowout-prevention device atop it. Similarly, these regulations called for approximately 870 feet of *surface casing*--a secondary string set to greater depth and generally installed when exploratory operations suggest the presence of a high-pressure gas-zone. Yet Solanas--exercising his legitimate statutory discretion--had authorized Union to drill A-21 without installing any surface casing at all. Moreover, he had permitted Union to run its conductor casing down to only 238 feet beneath the ocean floor.

Platform A had promised more for less, but such was not to be. By early afternoon of January 28 Union Oil Company management sensed the clear and imminent danger of a petroleum Vesuvius. Having reached bottom early that morning, the drilling crew proceeded to remove the drill pipe, stand by stand. At 10:45 a.m., as the crew was hoisting out the eighth 90-foot stand, mud began to trickle up inside the hollow drill pipe. The trickle turned quickly into a geyser of mud followed by roaring gas. Almost momentarily, the platform deck and crew were drenched with torrents of mud while the atmosphere became filled with hydrocarbon mist.

cities of Santa Barbara and Carpinteria, had joined
forces to file suit against Union[6] for $560,006,000
damages to public property. By early February, class
suits had been filed in federal court in the name of
four groups--local property owners, the fishing industry,
other individuals connected with boating, and finally
those with aesthetic interests in using Santa Barbara
beaches.[7] Individual suits proliferated until they virtually defied enumeration.[8] When faced with the prospect of stiffer and more expensive drilling regulations,
some companies requested that the federal government return the funds they had paid for leases in the Channel.[9]
Indeed, on April 9, 1969, six minor producers, headed by
Pauley Petroleum, filed a petition in the United States
Court of Claims seeking recovery from the federal government of lease, exploration, and development costs on two
parcels, Tracts 375 and 384, on precisely such grounds.[10]

On the "criminal front," Santa Barbara District
Attorney David Minier argued in state court that Union's
spill constituted a public nuisance. Such a misdemanor
finding would result in a penalty of up to $500 for each
day that the nuisance condition continued thereafter.[11]
Simultaneously, Santa Barbara City and Santa Barbara
County sought a permanent injunction in federal courts
against all further drilling off the California coast on
the grounds that any monetary damages secured for the
existing pollution would not constitute an adequate
remedy.[12] Finally, and most grandly, they sought a
larger--and some might think, last--constitutional
resort. In the same suit they asked for a declaration
that the 1953 Outer Continental Shelf Lands Act's[13]
delegation of leasing power to the Secretary of the Interior was so vague and unconfined as to violate the
due process clause of the Fifth Amendment.[14]

By Midsummer Day, 1969, sufficient political pecu-
liarities had surfaced to raise the possibility that
the Santa Barbara oil-slick controversy would be something more than the tale oft-told in the American past
of "restless natives"[15] vainly seeking to obstruct the

"progress" of industrial development in the name of
agrarian tradition. Speaking for the Rhode Island
Supreme Court as late as 1934, Judge Murdock had summed
up with unusual "judicial realism" for his day the
policy clash involved in this older order of conflictual
things. Turning away a farmer's complaint that Socony-
Vacuum's method of disposing of its refinery by-products
was rendering his water well unusable, Judge Murdock had
asked:

> A query arises as to whether the
> divergence of views expressed in
> these cases is not due to the influ-
> ence of the predominating economic
> interests of the jurisdictions to
> which these apply; in other words,
> whether these opinions do not rest
> on public policy rather than legal
> theory....It will be observed that
> in jurisdictions holding that, even
> though there is no negligence, there
> is liability for the pollution of
> subterranean waters, the predomin-
> ating economic interest is agri-
> cultural.

Here, however, he went on to observe:

> ...defendant's refinery is located at
> the head of Narragansett Bay, in the
> heart of a region highly developed
> industrially. Here it prepares for
> use and distributes a product which
> has become one of the prime necessi-
> ties of population, and its segrega-
> tion in restricted areas that indi-
> vidual rights recognized in a sparsely
> settled state...[has] to be surrendered
> for the benefit of the community as
> it develops and expands. If, in the
> process of refining petroleum, injury

is occasioned to those in the vicinity,
not through negligence or lack of
skill or the invasion of a recognized
legal right, but by the contamination
of percolating waters whose courses
are not known, we think that public
policy justifies a determination that
such injury is damnum absque injuria.
[An injury without a legally recognizable wrong.][16]

PERSISTENCE AND SUPPORT

The first 6 months of reaction suggested strongly
that the Santa Barbara gnat would not accept such a
view and give up. Beyond the "normal" annoyances to the
oil industry of law suits and petitions to "Get Oil Out,"
were nastier implications. First, the local and normally
moderate Santa Barbara *News-Press*, "The Oldest Daily
Newspaper in Southern California," was pursuing the matter in virtually every issue with a depth of coverage[17]
unmatched since its "pioneering" and Pulitzer-Prize-
winning investigations of the John Birch Society almost
a decade earlier.

Second, the political sensibilities aroused by the
spill impelled California's liberal freshman Senator,
Democrat Alan Cranston,[18] to take action, and his efforts
were easily matched by the three bills introduced by
Congressman Charles Teague.[19] The latter was a six-term
Republican veteran not generally given to espousing liberal causes with potentially antibusiness implications.
By February 5, 1969, however, Teague was ready to join
the Santa Barbara City Council and the County Board of
Supervisors in seeking to bar not only future federal
drilling but present production in the Channel. This
fact was all the more intriguing, since a majority of
Teague's constituency[20] resided in Santa Barbara County's
southerly neighbor, Ventura County, where attitudes were
ambiguous on the issue.[21]

Third, on March 7, 1969, Orin E. Atkins, president of the minor nonintegrated Ashland Oil and Refining Company, was quoted as saying that several companies might not be altogether sorry to see the leases rescinded--assuming a *quid pro quo* from the government.[22] In the words of Byron E. Calame, reporter for the *Wall Street Journal*, the oil companies had been "tapping little but trouble."[23]

Fourth, former Secretary of the Interior, Stewart Udall, under whose aegis the lease bidding had been opened despite abundant local protests in 1967, declared on March 10, 1969, that he had been mistaken. He called the Channel leasing a "conservation Bay of Pigs."[24] Finally, the American Civil Liberties Union agreed to take up Santa Barbara's cause, and thus commenced its first legal battle on behalf of environmental rights.

At the very least, local outcry against the pollution had made the oil companies uneasy in Santa Barbara. Union Oil Company was sufficiently unsettled that at one point it considered moving a land base of its supply operations from Stearns Wharf in Santa Barbara harbor to Port Hueneme in Ventura County, despite the fact that shore-to-platform sailing time would be tripled. Stearns Wharf had earlier been the scene of a confrontation between unhappy citizens and oil company workmen delivering supplies. And in June 1969, the Santa Barbara County Board of Supervisors denied a permit to Sun Oil Company to run a pipeline from a proposed platform through Santa Barbara County waters to a proposed onshore facility in Ventura County.[26]

To be sure, none of these facts suggested that the oil companies would be forced to conform to native wishes immediately. Neither had local clamor two years earlier dissuaded the federal government from opening the Channel for leasing: In 1967 numerous local conservation groups, spearheaded by the Sierra Club's Los Padres chapter, had objected vociferously to leasing without consultation with the community.[27] Yet all they secured from the

federal government was a brief hearing on the construction of one platform by Phillips Petroleum which the
Army Corps of Engineers, after months of pressure, had
finally consented to schedule in Santa Barbara on November 20, 1967. The empty formality of that hearing was
emphasized by the facts that oil industry representatives
had largely crowded out local citizens and that, as it
transpired, Phillips' platform was already enroute to
its Channel site via barge.[28]

Conservationists had scored a greater--but very
temporary--success in May 1967 by motivating the Santa
Barbara County Board of Supervisors to ask the Interior
Department for a one-year moratorium in order to undertake a study of the likely "local fall-out." Additionally, the board had requested that Congress extend the
three-mile-wide marine sanctuary granted in 1955 by the
State of California.[29] Yet trips to Washington, led by
the most active supervisor, George Clyde, representing
the wealthy southeasterly suburb of Montecito, went
largely unheeded. The best that the county could obtain
to allow time to prepare a report, was a 60-day moratorium (in September 1967),[30] a federal "no drilling"
buffer zone extending two miles out from the California
three-mile limit, and assurances that the platforms
would be camouflaged by paint and made as aesthetically
pleasing as possible.[31]

Admittedly the Santa Barbara *News-Press* may have
been overly dramatic in insisting that the oil spill
marked a potential turning point, and perhaps the final
one, for reappraising the national interest in allocating
priorities between conservation and exploitation. Nevertheless, the initial months did present an undeniably
peculiar phenomenon--a strange cleavage that cut across
normal class, interest group, and party alignments.

Outlines of the Analysis

Virtually all parties to the conflict agree on one point: the spill is a standard side effect of current patterns in American natural resource development. But this side effect also raises at least four basic questions that help shape the outlines of this monograph's analysis.

First, does the side effect fall upon a rural backwater and thus implicitly not deserve singular analytic attention? Or, does it represent spillovers from modern economic and living patterns such as those engendered by megalopolis? If the latter is the case, then the clash of interests clearly warrants the examination it receives in Chapter II: "The Santa Barbara Economy: Agrarian Remnant or Post Megalopolis."

Second, does the side effect create minor ecological and financial losses or burdens that should be allowed to fall where they may? Or, alternatively does it create heavier burdens that should be averted, apportioned, or transferred? See Chapter III: "How Heavy Is the Local Burden?"

Third, is the side effect only a "local harm," perhaps one that should be overridden by national and federal interests? Or, does its "apportionment" require reappraisal? See Chapter IV, "Offshore Drilling and the Public Interest." Initially, we must ask whether any present offshore drilling is in the long-term national interest and, if so, where? To what extent might alternate sources of petroleum suffice--particularly Colorado's oil-bearing shale and recently discovered reserves in Alaska? Beyond the possibilities of alternate sources are two major technological questions requiring consideration. First, perhaps certain kinds of offshore geological formations are indeed safe enough for drilling near a residential community, and perhaps others are not. For that reason it is essential to

examine the geology of the Santa Barbara Channel to determine whether its features are so unusual as to require special caution. Next is it important to consider
what is likely to happen after a blowout? A relatively
sure cure for a blowout--that is to say, technology permitting damage to be localized before it can affect a
broad oceanic or littoral area--would require a very
different weighing of the values and dangers, both economic and ecological. To this end, Union Oil Company's
remedial measures are examined as a case study of the
adequacy of present technological capabilities for cure.

Fourth, does the nature of the problem posed by the
Santa Barbara oil slick, and more generally by offshore
drilling using existing technology, pose the kinds of
conflicts that can be resolved satisfactorily under established contemporary pluralist patterns of "articulating and aggregating"[32] competing interests? These
pluralist patterns are manifested in overlapping jurisdictions among federal agencies, each agency being sup-
ported by its own clientele. They are also found in
federal-state-local divisions of authority and jurisdiction, each level having its particular constituency to
which it must respond. Three chapters of the monograph
attempt to explore this question. Chapter V suggests
that the federal-state relationships concerning oil and
pollution have developed by accretion, without any
clearly thought-out plan. Chapter VI outlines the executive and legislative remedies apparently now available
in a "Topsy-like" situation, while Chapter VII appraises
the lay of the land confronting those who seek judicial
redress of grievances.

A pessimist might conclude that the Santa Barbara
oil conflict is not likely to be resolved by swift and
firm application of "neutral principles"[33] to the arguments put forward by the contending parties. Rather,
he might expect that the ideological axioms of "Interest-
Group Liberalism"[34] would guide the conflict into "government by bargaining."[35] He might also anticipate that,
because the bargainers are so unequally matched, a decision in favor of the oil industry would be inevitable.

Should the opposite result come about, he would probably attribute the outcome to a "fluke of personality," for instance, a conservation buff happening to be in a pivotal governmental position at the right time, rather than giving credit to the normal workings of the American political process for the "favorable" end result. Accordingly the concluding chapter of this essay seeks to determine whether such pessimism is fully realistic in the light of the recent and rapid emergence of ecological consciousness in the American polity.

CHAPTER II

The Santa Barbara Economy:
Agrarian Remnant or Post Megalopolis

THE OIL INDUSTRY'S PERSPECTIVE

From the oil industry's perspective, President Fred Hartley was doubtlessly correct on one point as he testified before Senator Edmund Muskie's Public Works Subcommittee on Air and Water Pollution in Washington on February 5, 1969. It might be true that numerous birds were killed by the oil slick, but to the oil industry that fact alone was insufficient reason to view the situation as a grand-scale tragedy. Nor was it made such by the interested citizens who worked ardently at local treatment centers to clean oil off 1,500 afflicted birds, the vast majority of which died anyhow. Nor was it clear even when 150 sea lion pups and 5 whales were found dead on beaches during the following months that the continuing spill justified calling a halt to attempts to gain a return on the nearly $1 billion the oil companies had already invested.

Furthermore, it might indeed be somewhat inconvenient for the inhabitants to have to use more gasoline than usual to remove tar from their feet after swimming or strolling on the beach. Still, the presence of scattered tar patches on the local beaches had long been part of Santa Barbara community life. Father Font, a Spanish missionary, had noted in 1776: "tar which the sea throws up is found on the shores, sticking to the stones....Perhaps there are springs of it which flow out into the sea."[1] In prehistoric times the Chumash Indians used the natural tar to bind their woven baskets;

later, Spanish explorers took advantage of its presence
to caulk leaky ships. During the latter half of the
nineteenth century, considerable profit was reaped from
seeping natural asphalt by at least a few enterprising
Santa Barbara citizens who shipped it north to pave
sidewalks and streets in Victorian San Francisco.

Similarly, the first discovery of substantial quantities of oil in the county brought less than distress.
In 1895 the discovery of oil, while drilling for water
near the beach, brought to a hasty end the "other-worldly"
aspirations of the inhabitants of Summerland, a coastal
hamlet four miles east of Santa Barbara just established
as the western home of the Spiritualist movement. By
1901 approximately 350 small wells had been drilled.
Santa Barbara did not then object to being the world's
first site for offshore oil development. Indeed, no
one seemed to bother too much about the failure to cap
most of the wells after their abandonment by the end of
the 1920s. Nor did anyone object to the May 1929 discovery of an oil field located less than a mile from
the heart of the town. If feelings of distress prevailed shortly thereafter, they stemmed from the unprof-
itability of money invested: Seepage of salt water
through faults in the oil sands made its separation from
crude petroleum economically prohibitive. Finally, no
serious objections had been voiced in 1928 following the
discovery of an oil field at Ellwood, 10 miles west
along the coast--an oil field that produced 99 million
barrels of oil before it played out.[2] Surely then, Santa
Barbarans' attempts to talk in terms of "vested aesthetic
rights" against Channel oil development came with ill
grace when induced by only a single blowout in 40 years.

AN ALTERNATE PERSPECTIVE

Taken individually, all the foregoing propositions,
except perhaps the conclusion, are unassailable. But a
counter argument challenging the oil industry view can
be based on more recent development of Santa Barbara's

socioeconomic structure. During the middle third of the
twentieth century that structure changed from a traditional, dominantly agricultural economy to an amenity-
oriented economy in which aesthetic considerations became vital.

By the Roosevelt era Santa Barbara was well established as a resort community and--by virtue of many
eastern vacationers who decided to stay permanently--as
a retirement center. By World War II Santa Barbara's
greatest previous disaster--a devastating 1925 earth-
quake--had placed the city in the forefront of communities concerned with environmental quality in the developing American post-industrial society. Following the
quake, a group now entitled The Community Arts Association offered free architectural advice to owners of de-
molished businesses who agreed to rebuild in a graceful
Spanish style. At the same time, city ordinances that
restricted advertising signs and required landscaping
kept Santa Barbara from developing into the typical
California urban sprawl of gas stations, supermarkets
and taco stands. In a similar vein the affluent suburb
of Montecito gave land to the state to develop a freeway
upon the condition that the state prevent the growth of
a commercial strip alongside it. The state agreed to
undertake the extensive landscaping of live oaks and
palms which makes the eastern entrance to Santa Barbara
famous.

By the middle 1950s a second major change had taken
place. General Electric's erection of its Tempo Research
and Development Center started what promised to be a
welcome flood of light industry, whose side effects did
not threaten established retirement and resort sectors.
Finally, the conversion of Santa Barbara College in 1958
into a full-fledged University of California campus
opened the possibility of a fourth major sector to the
service-oriented economy--higher education.

As Table I shows, projections made by the county in
1962 indicated a continuing shift in Santa Barbara
County's "South Coast" economy away from agriculture.

TABLE I

SANTA BARBARA COUNTY AND SOUTH COAST INCOME
(In Millions of Dollars)

	Santa Barbara South Coast		Entire County	
	1960	1978*	1960	1978*
Missile Bases	$ 4.2	$ 4.4	$60.0	$80.0
Visitors	23.2	60.7	30.2	86.7
UCSB	6.4	36.3	6.4	37.0
Manufacturing & R & D	19.7	50.0	21.7	71.5
Property & Pensions	29.2	53.4	33.6	66.7
Agriculture	12.7	6.3	40.4	57.0
Mining	2.2	2.5	7.5	10.1
Other	7.6	13.6	9.0	20.0
Totals	105.2	227.2	214.8	429.0

*Calculated in 1960 dollars

The university, the research and development industry, and the resort and retirement sectors have all grown tremendously from their 1960 base, while agriculture is expected to decline to less than 3 percent of the South Coast economy by 1978. In 1962 it was anticipated that mining would remain at about 2.5 percent.

Table I also discloses the very different types of development projected for the South Coast and the northern parts of the county. Geographically cut off by the California Coastal Range, the "North County" is committed to growth in agriculture, mining, manufacturing and missile bases. To note this is to point to a crucial subcondition of this counterargument: the division of interests between the northern region and the South Coast.

If all these premises are accepted, the Santa Barbara South Coast's economy seems best understood as a forerunner of a late-twentieth-century economy, which would leave behind both traditional agricultural structures and heavy-industrial metropolitan ones. On this showing the resemblance of Santa Barbara's natural evolution to modern planned communities of the future such as Columbia, Maryland, constitutes strong ground for reexamining the nature of the conflict between new and old.

CHAPTER III

How Heavy the Local Burden?

DIFFICULTIES OF "MEASUREMENT"

It is difficult to deny that, even without a continuing oil spill, offshore oil production imposes an appreciable aesthetic burden upon a community with Santa Barbara's socioeconomic and ecological structure. But, two points are open to dispute. First, in the absence of an oil spill, to what extent is the burden offset by oil production's direct and indirect monetary infusion into local coffers--both public and private? Second, how much is added to the debit side of the ledger by an oil spill--in both short- and long-run terms?

It is well to note that three types of difficulties are involved in measuring the net burden. First, the analyst is necessarily faced with measuring incommensurables, like the value of having chickens' eggs that don't crack when boiled, as opposed to the value of having a car that accelerates to 60 m.p.h. in eight seconds flat. How much does aesthetic discomfort offset, for each citizen, the advantage of a possible reduction in his property taxes due to the infusion of taxable oil revenues into the community? And how is such a reduction balanced against a possible reduction in the market value of his home? And further, how does one compute an average for the entire community? Second, since much information is not open to public scrutiny but secured in the files of the Department of the Interior and the several oil companies, the analyst's access to potentially relevant data for measurement is

18

necessarily incomplete. Third, even if he were to come
into possession of all available information, it would
still be difficult to predict future technological developments in relation to prevention and cure. Past
spills--such as the *Torrey Canyon* incident off the
Cornish Coast of Great Britain--have not provided data
reliable enough to make definite long-range projections.[1]
With these caveats in mind, what can be said?

DIRECT AND INDIRECT ECONOMIC EFFECTS

What direct advantage does the South Coast of Santa
Barbara receive in purely fiscal terms? For 1967-68,
the oil companies' contribution in tax money was less
than 2 percent of the total revenue.[2] Indirectly, the
fiscal contribution may have been somewhat greater.
The oil companies argue that, if their development
plans continue, well drilling, pipeline laying and construction of onshore facilities would bring hundreds of
new jobs with multiplier effects into the local economy.
Historically, so the argument continues, satellite industries have followed in the wake of the oil industry
and established themselves in nearby communities.

There are, however, three potential quarrels with
such an argument. First, the additional jobs and pay-
roll money may be primarily a short-run phenomenon.
Once drilling is finished and production well under way,
the number of oil employees can be expected to drop
sharply off from the maximum number employed during exploration and drilling. Second, the satellite-industry
argument may be less applicable here than in other, tra-
ditionally rural areas, such as west Texas. It is
doubtful that such satellites would make up for likely
dropoffs in anticipated new light industries and new
senior citizen residents, the power of whose attraction
has lain in the unspoiled natural splendor of the environment.[3] Third, additional population creates not only
economic benefits but also economic costs in terms of
public services such as schools, streets, police and
fire protection and water supply. Since the oil

installations add virtually nothing directly to the tax base of the city, the net result of oil development might be a serious burden to the public sector.

Given these doubts, one might well argue for a consideration of local public attitudes as to the benefits and disadvantages of Channel development--prior to the spill. Unfortunately, no reliable samples of relevant public opinion evaluating community attitudes were taken then, unless one wishes to repose full confidence in the Western Oil and Gas Association's finding that its 1967 poll of Santa Barbara County adult males disclosed 20 percent opposed to offshore oil developments, 17 percent in favor, and 63 percent indifferent.[4] Aside from doubts created by the poll taker's vested interest in the results, it is open to the criticism that, by sampling countywide, it ignored the crucial issue--the separate nature of the South Coast economy. However, the fact that less than one-fifth of the sampled population replied positively indicated little enchantment with the prospect of an oil boom. Perhaps more reliable evidence is found in a November 1968 referendum on a proposal to permit Humble Oil Company to construct an onshore oil processing plant. Voters in Santa Barbara County overturned the Board of Supervisors' decision to permit such a facility by a vote of 44,290 to 41,404.[5] While the issue was not entirely clear-cut, the results suggested considerable hostility to oil, whatever its manifestations.

Once the spill had occurred, however, little doubt remained. Communications flowed in great number to Washington,[6] far more letters were received by the Santa Barbara *News-Press* than on any other prior issue; and Congressman Teague concluded from his mail that Santa Barbara South Coast residents were "virtually unanimous in supporting ceasing [*sic*] of all oil activity in the channel."[7] Within a month after the spill an estimated 70 percent of the adult South Coast population had signed the GOO Committee petitions seeking to "Get Oil Out," and during summer and fall of 1969, GOO's petitioning efforts continued successfully. By

early January 1970, the executive head of the GOO organ-
ization, Lois Sidenberg, reported that 200,000 signatures had been obtained on a petition seeking an end to
Channel drilling and production.[8] By March 1970, active
GOO membership had doubled in excess of 2,000,[10] while
the January 28 Committee, originally organized to draw
national attention to the first anniversary of the
spill and to its continuance, reorganized itself on a
permanent basis as the Community Environmental Council.[10]
On Earth Day, April 22, 1970, the White House was sent
the petition of some 200,000 signatures from GOO de-
manding a complete cessation of Channel oil development. Apparently unsure just how to handle this dubious gift, the White House sent it to the Department of
the Interior two weeks later.[11]

ECOLOGICAL DAMAGE

One of the major causes of Santa Barbara's hostility to Channel oil development, and the one most immediately apparent after the spill, was fear of damage to
the natural environment. One factor that substantially
contributed to both fear and hostility was the apparent
tendency of the federal government and the oil industry
to make the situation seem less serious by consistently
giving Santa Barbara citizens apparently low estimates
of the amount of oil spilled. Undeniably, it was difficult for any analyst, public or private, to give exact
figures describing the rate at which the Channel was
being polluted by the blowout. In view of the damage
and the wide geographic dispersion of oil upon the
ocean's surface, it was difficult for many Santa Barbara eye witnesses to accept governmental estimates
which ran consistently five to ten times lower than
those made with much greater care by Allen A. Allen,
an executive of General Research Corporation, a Santa
Barbara research and development firm. Nor was the
government's credibility gap diminished by its practice
of scaling upward its earlier estimates as time passed.[12]
Consequently, sharing these local doubts, we place more
confidence in Allen's estimates based upon a comprehensive sampling and scaling of the thickness of oil on the

22

water.[13] Allen's figures have the obvious advantage of
not being so belied by posterior events as to require
substantial readjustment in order to achieve even a
plausible patina of truth.

Allen estimated that between 420,000 gallons and
2.1 million gallons were spilled during the first two
days of the flow. By February 3, six days after the
blowout, oil had spread over 251 square miles of the
Channel. Eighty-six square miles were covered with
heavy, dark oil, the remainder by a lighter film. By
February 5, Santa Barbara County beaches were blanketed
with a layer of crude oil which was several inches
thick in many places. Its odor was noticeable several
miles inland; 660 square miles of the Channel were cov-
ered--160 square miles by the heavy, dark oil. (See
Figure II,)

The flow of oil was rapid for the first 11 days of
the spill. During the next month, the flow began to
taper off, down to an estimated 21,000 gallons per day
by the end of March. In April the flow diminished further to somewhere between 4,200 and 8,400 gallons per
day; it so continued into summer 1969.[14] During late
summer and early fall the rate declined again--down to
what appears to be a fairly constant flow of about 420
gallons per day,[15] with occasional greater spurts induced by minor seismic activity in the Channel.[16] In
sum, the best estimate is that by the end of April 1969
at least 3 million gallons had been spilled in the Channel. As Figure II shows, the geographic dispersion of
the oil on the ocean's surface was extensive. Oil was
reported as far away as Pismo Beach, 90 miles to the
northwest, and Malibu Beach, 65 miles to the southeast.

The Santa Barbara oil spill was at the time the
third largest in the oil industry's offshore history
(see Table II). Since then, three blowouts off the
Louisiana coast have exceeded it--those of Chevron and
Shell in 1970, and American Oil in 1971. Probably because
Louisiana derives 40 percent of its state revenues from
the oil industry, the outcry has been neither so loud

Within five days after the Santa Barbara spill, tidal pools and beaches were covered with sheets of oil. Hundreds of birds became covered with the sticky mess and began to die unpleasant deaths. Almost immediately, three treatment centers were set up by the local residents to care for oil-soaked birds. During the first month after the spill 1,575 birds were brought in for treatment. Initial estimates were that 80 percent of these birds died,[18] and that the total number of birds killed, those treated and untreated, was as high as 8,000. Officials pointed out that a major avian disaster had been averted because the area's bird population happened to be unusually low at the time of the spill. Being a major stopping point on the Pacific north-south flyway, Santa Barbara might have witnessed the extermination of an entire species if one of the bird types that migrate en masse, such as the brant, had been passing through at the time.[19] At first, Santa Barbara citizens were able to take somewhat cold comfort in evidence that they had managed almost twice the saving rate of those Cornishmen who had treated oil-soaked birds after the *Torrey Canyon* disaster. A year later, however, the California State Fisheries and Games Department reported that this appearance of relative success was illusory. Although estimates of the total bird population killed came down to 3,600, the survival rate of those treated declined to under 11 percent.[20]

Immediately after the spill, general assessments of the damage done to the Channel's ecology ranged from a petroleum industry scientist's contention that little damage had been done to wildlife to alarmist opinions that the Channel's ecosystem had sustained irreversible damage. During late spring and early summer, a particularly sharp dispute arose between governmental agencies and private analysts whether the death of five whales whose portly cadavers washed up on various California beaches was related to the spill, and whether some 150 sea lion pups found dead on San Miguel Island had died because of Union's mistake. *Life* magazine argued affirmatively and contended that the oil drilling seriously jeopardized the future of sea lions in the

Channel. The Department of the Interior insisted that
there was no evidence for such conclusions.[21]

　　While it is still not possible to be altogether
certain about the extent of ecological damage sustained,
a year and a half after the spill the fairest estimate
seemed to be that the result was a mitigated disaster--
mitigated largely by the presence of giant kelp beds
that run the length of the Santa Barbara Channel a few
hundred feet seaward of the beach. A study financed
primarily by the Federal Water Pollution Control Administration and executed under the direction of Professor
Michael Neushul of the Department of Biological Sciences
at the University of California, Santa Barbara, found
that heavy biological damage had occurred in the intertidal areas to both the barnacle population and to the
surf grass. The latter was still being affected by the
continuing spill, whereas the giant kelp, apparently
protected by its thin natural coating of mucilage, was
not significantly damaged. Like the kelp, the common
intertidal anemone appeared highly resistant even to
heavy localized concentrations of oil. Ecological damage in the *Torrey Canyon* incident was attributed after
the fact to the use of detergents in emulsifying the
oil. Paradoxically, Union Oil's use of a water-and-
sand mixture, pumped under high pressure to clean oil-
covered rocks, eliminated a large number of crabs,
algae, limpets and snails.

　　Reflecting the findings of some environmentalists,
Dr. Dale Straughan from the University of Southern California's Allan Hancock Foundation states that her research

> ...failed to reveal any effects of oil
> pollution on the channel's zooplankton
> and phytoplankton; similarly, sea
> plants and the production of fish and
> larvae were not lastingly affected.
> Although one variety of barnacle was
> smothered by the oil...other species--
> including local seals and the migrat-
> ing gray whale--escaped unharmed.[22]

Other environmentalists disagreed with these findings. Dr. Kristian Fauchald who worked with Dr.
Straughan on the project, disagreed so strongly with her
findings that Dr. Straughan reportedly screamed at him,
"I did not sell out to the oil companies."[23]

Perhaps the most careful scientific study on the
general chemical and biologic effects of oil spills has
been conducted by specialists at the Woods Hole Oceano-
graphic Institution. They are of the opinion that many
of the statistical studies, such as those conducted in
the Santa Barbara Channel are far too insensitive to
have much value and that their studies indicate that
oil pollution has grave effects on the marine environment.[24]

BUSINESS AND RESIDENTIAL PROPERTY DAMAGES

The oil spill's long-run effects upon the fishing
industry remain uncertain, but there is little doubt
that catches declined soon after the spill. Aerial
fish spotters reported sighting many large schools of
fish during January but virtually none during February.
Causal attribution of this decline, however, is rendered difficult by the circumstance that abnormally
large rains in late January and early February deposited
large quantities of silt in the ocean from flooded
streams and rivers. It is just possible that the silt
was responsible for the apparent dearth of aquatic life.

Interviews with fishermen evoked complaints of
great nuisance to boats and netting equipment caused by
the oil, but they were not unanimous in attributing the
poor fishing to the oil. One fisherman said optimisti-
cally, "fish don't swim on top of the water with their
mouths open." Others were more concerned that the effects of the disaster could not be foreseen. They
argued that the chain of forage fish may have been
greatly damaged, and the effect of the spill on fishing
would take years to assess. In point of fact, the
total fish tonnage landed at Santa Barbara during 1969

was the highest in five years. However, as late as the
end of January, 1970, local fishermen were predicting
trouble ahead, some noting the absence of whole species,
such as halibut, and attributing this to a lack of the
halibut's normal food--plankton, anchovies and sardines.

Boatowners and businesses related to boating were
among the first to feel damage from the spill. Every
boat in the Santa Barbara Harbor received a substantial
coating of oil. Union promised that their insurance
companies would pay for cleaning the boats and one
month's slip rental.

However, Union's insurer was discovered to have
inserted a fine-print waiver of further damage suits in
the boat-cleaning claims form.[25] Ultimately, Union decided not to contest liability for boat cleaning in
court. About a year and a half later 106 boat owners
settled for $105,325.72. Attorneys for 30 other boat-
owners, however, dubbed this a sellout and determined
to press on with suit.[26]

Businesses sustaining heavy losses from oil damage
in the harbor area included boat brokers, some restaur-
ants near the shore, charter fishing boats, boat rental
companies, and marine and fishing supply companies. One
of the supply companies reported that February 1969
was its worst month in eight years. Charter fishing
and boat rentals were at a standstill and one large
restaurant reported business off 50 percent.

Motels in Santa Barbara account for approximately
one-fifth of the community's tourist income. A survey
of 29 Santa Barbara motel owners in early March 1969,
disclosed that up to that point direct oil damages from
the spill appeared minor. However, questions concerning long-term losses uncovered anxieties that the oil
spill would have a major damaging impact on future
business. One spokesman for the motel owners said the
industry was very concerned about long-range losses
which might be caused by adverse oil spill publicity
and that the oil companies had already been notified

through counsel that they would be held responsible for such losses. Overall estimates of future annual losses by motel owners ranged from no losses at all to $3.5 million.

By early summer of 1969 the pessimists seemed to have been the better prognosticators. For the first time in many years, beachfront motels displayed vacancy signs in late June.[27] The Santa Barbara City bed tax-- calculated monthly according to the number of rooms rented in hotels, motels, and boarding houses--was off from the preceding year's intake by 8 percent in June, 12 percent in July, and 5½ percent in August.[28] Over the long run it appears that future losses, if any, are more likely to be sustained by beachfront motels and hotels than by those in other parts of Santa Barbara whose business depends less on resort traffic and more on the volume of north-south highway traffic stopping overnight and on the number of persons visiting Santa Barbara on business.

Indeed, economists Walter J. Mead and Philip E. Sorensen noted that tourism for the City of Santa Barbara proper declined during 1969 but that the county as a whole showed an increased income from tourist oriented businesses. This resulted from a dramatic increase in tourist revenues in nearby Goleta Valley. Thus Mead and Sorensen concluded that there was no net damage to tourism in the area, though beachfront motel and restaurant owners may have suffered. Nonetheless, the fact that the tourist business was good elsewhere obviously afforded little satisfaction to beachfront motel and restaurant owners.

Mead and Sorensen's interviews with real estate dealers knowledgeable in beach front property sales indicated a strong consensus that:

(1) the volume of beachfront property sales declined sharply from 1968 to 1969 and 1970;

(2) market values suffered a short-term decline in the range of 15-25 percent due to the oil spill;

(3) about 250 beachfront properties were affected
in the region from Montecito to Rincon.[29]

Questionnaires distributed to a sample of 35
beachfront homeowners indicated that immediate direct
damages by oil to seawalls, fences, gardens and resi-
dences exceeded $1,000 for many owners. Indirect losses
mentioned far exceeded this amount. One respondent
claimed a loss of $1,500 a month in rent and another
claimed a $5,000 depreciation in the value of his home.
The residents reported that they had paid extremely
high prices for their property, as much as $1,000 per
beachfront foot. One lady reported that she was compelled to move out of her home because of fumes from
the oil. While other owners were not so extreme in
their response, they were being denied use of the
beach--for which they had paid such high sums in the
purchase price. Perhaps the general attitudes of these
affluent property owners was best summed up by the concluding query of one respondent: "Know where we can
get a Stanley Steamer?" or by two other remarks, extraordinary for their coming from a bastion of wealthy
Southern California Republicanism: "...complete loss of
confidence in the federal government and *Big Business*;"[30]
and "The federal government is basically responsible....
No taxation without representation. No oil company
should have unlicensed power to contaminate the English,
French, and California coasts."[31]

These affluent residents of Montecito caught the
general sense of the larger Santa Barbara community,
and, these attitudes had not greatly softened many
months later. In early December 1969, after a week and
a half of continuously high surf, substantial oil
washed ashore ten miles east of Santa Barbara at Carpinteria State Park Beach. In addition, Union oil managed to stage a repeat performance with another accident
on December 16, 1969. On that date a pipeline broke on
Platform A, forcing Union to stop the operation of all
30 wells on that platform. During the four days required to repair this new "difficulty," 60 grebes were
found covered with oil and dying, and a heavy coating

of crude oil washed ashore at Rincon Beach, one of the two or three best winter surfing spots on the Southern California coast.[32] A few days later Union Oil President Fred Hartley announced that his company sought to build another platform (B) in the area of Platform A as soon as possible.[33] Moreover, once that later platform went into operation Union and its partners sought unsuccessfully to gain permission to erect a third one during 1971.

Finally, more than two and one-half years after the spill, Santa Barbara's suspicions were rekindled that Union or its subcontractors were still not practising oil-exploitation methods that maximized existing technological capacities for preventing further disasters. In the Fall of 1971 a former member of the Platform A drill-crew, William J. Gesner, charged that Union's drilling contractor, Peter Bawden Drilling, was continuing to ignore federal law governing safe drilling and extraction practises.[34] While Gesner's contentions have not been proved in court, neither has Union successfully refuted them.

CHAPTER IV

Offshore Drilling and the Public Interest

The concept of the public interest is as slippery
as oil itself. A workable definition must specify the
relevant public, the generality and intensity of their
interest, and the manner in which that interest is mea-
sured. Seldom is this done with any precision. Its
utilization in political struggles tends to be as a
weapon rather than as a scientific explanation. We attempt here a dispassionate appraisal.

THE ECONOMICS OF OIL AND THE QUESTION
OF ALTERNATE SOURCES

The principal arguments of the oil companies and
of those federal officials who defend offshore oil development in Santa Barbara emphasize the economic and
strategic aspects of the national public interest which
such development purportedly promotes.

The oil companies stress that an oil-dependent
economy needs to develop additional productive oil resources. Even when urging that the oil industry must
control its own pollution "regardless of the extra cost
or how that cost is absorbed,"[1] oil industry leaders
seem to assume without question that "fossil fuels are
and...will continue to be, the principal source of
energy for the remainder of this century."[2] Consequently, there is little room in the oil industry's
scheme of things for the point of view that it is most
unwise to be pumping out and using up our exhaustible
fossil fuels simply to provide sources of energy.[3]
There are persuasive arguments that we should be developing other sources of energy as rapidly as possible
and retaining earth's irreplaceable supply of fossil
fuels for use as chemical building blocks to synthesize

the vast variety of substances and materials that civi-
lization may need urgently in the near and long-term
future.

The Oil Industry Perspective

From the perspective of the oil industry, however,
the country's needs can be met only by rapid development
of domestic and foreign oil sources. In this view, the
national security of the United States depends on the
continued exploration and development of domestic resources because of the volatility of the political and
military situation abroad, particularly in the Middle
East. With proven oil resources becoming more and more
difficult to exploit in the United States, the oil companies see the necessity of developing new potentially
productive areas to meet increasing needs.

The federal government, and presumably taxpayers
in general, have a stake in offshore development.
Awarding the Channel leases immediately brought over
$600 million into the United States Treasury and unspec-
ified future amounts in the form of royalties of 16.67
percent of the value of crude production are promised
for the future. To a government hard pressed to find
revenues to fight both a war in Southeast Asia and a
war on poverty, such financial resources had an obvious
appeal. Moreover, it appears that the federal officials
saw oil investment in the Santa Barbara Channel as a
desirable alternative to greater foreign imports, which
detrimentally affect the balance of payments of the
United States.[4]

Critics of Offshore Development

The critics of offshore oil development argue that
such development is not in the national interest, at
least not at the present time. One principal argument
emphasizes that only import tax and quota systems make
a high-risk area like the Santa Barbara Channel an

attractive proposition.[5] At the time of the spill, the
oil industry enjoyed a controversial depletion allowance that permitted oil companies to deduct, for tax
purposes, 27.5 percent of the gross income from their
producing property but not more than 50 percent of the
net income from that property. The new tax law passed
in December 1969, doubtless in part a result of public
reaction to the Santa Barbara oil spill, reduced the
depletion allowance to 22 percent. While the oil companies exhibited much discomfort and exerted unsuccessful pressure to prevent this reduction, and while industries' fear that the alteration represents the first
appearance of the camel's nose of equity under the tent
of special treatment may be justified in the long run,
there nevertheless remains sufficient tax incentive for
high-risk development. American producers abroad are
given additional tax advantages by being allowed to
deduct royalties paid to foreign governments from the
United States income taxes.[6] The combination of these
two advantages allowed the 22 largest American oil producers to pay only slightly more than 6 percent in federal taxes between 1962 and 1966; for most non-oil corporations the tax rate was just under 50 percent. As
an added incentive to "uneconomic development," in 1959
the federal government established oil import quotas
which guaranteed a percentage of the market to domestic
oil producers. The result is that the prices for oil
in the United States are usually substantially higher
than they are on the world market. At the time of the
spill, the price of oil on the world market was about
$2.00 per 42 gallon barrel, while the United States
price was over $3.00. Additionally, import quotas are
combined with market-demand pro-rationing among domestic producers, which further curtails competition and
maintains artificially high prices.

Production and Marginal Areas

Critics argue that these practices encourage industry to press for development of oil production in areas
that are presently marginal by existing technology. In

the absence of such practices, the oil industry would
be forced to operate only where costs were sufficiently
low to warrant it, and this would exclude the Santa
Barbara Channel under existing conditions. While this
argument does not necessarily preclude eventual oil
development in the Santa Barbara Channel, it suggests
postponement until demand favors such development and
until technology improves appreciably.

As if to recognize the force of this argument, in
February 1970 a majority of President Nixon's task force
recommended overhauling the system of oil import quotas
in favor of tariffs, which would have taken effect in
1971. At the time, the President's reaction was half-
hearted; he did not proceed beyond establishing a permanent Oil Policy Committee under the leadership of
General George A. Lincoln, Director of the Office of
Emergency Preparedness.[7] It is thus difficult not to
conclude that, in relation to the oil industry, laissez-
faire economics is an ideology, a publicly propounded
myth, rather than an operational reality.

Other Challenges to Development

A second and much more sweeping kind of attack upon
the oil industry's advocacy of Santa Barbara Channel
development is grounded in ecology rather than economics.
Three species of argument cluster within this attack.
The first questions the very basis of the oil companies'
assumption that rapid use of fossil fuels for producing
energy is advisable. This argument applies equally to
onshore and to offshore oil development. One such line
of argument, was mentioned before: that it is a mistake
to burn up oil resources as fuels rather than to con-
serve and develop them later as chemical building-blocks
for new materials. Another, perhaps slightly less futu-
ristic, also questions the basic values underlying a
high-consumption, high-growth economy: Population-
control expert Garrett Hardin has argued that the energy
needs of an increasing population in this country should
not be given high priority, since continued population

growth is a threat to the quality of American life.[8]
On this view of the matter there would be little reason
to worry if the oil resources in the Santa Barbara Channel are not developed to provide for the energy needs
of an additional several million Americans. Hardin argues, "Who needs them?" In Hardin's view the pollution
resulting from oil development would depress the quality
of American living far more than the energy produced
would improve it.[9]

A second line of argument is based upon the apparent difficulty which a "cowboy economy"[10] in transition
must face in building into its cost-benefit calculations
the unpleasant long-run consequences of unfettered high
consumption. This ecological argument is concerned
with the effect of oil pollution in the world's oceans.
Dr. Edward E. Goldberg of the Scripps Institute of
Oceanography in San Diego, addressing the Nuclear Science
Symposium of the Institution of Electrical and Elec-
tronics Engineers in October 1969, drew attention to the
fact that man is currently dumping in the ocean over
one million tons of oil per year.[11]

Most of the oil dumped directly into the ocean--as
distinct from that flowing in from rivers--currently
comes from discharges by ships. However, the prospects
for greatly increased discharges resulting from wide-
spread offshore oil development are certainly discon-
certing. Some scientists are distressed by the possible
effects which a very thin, micromillimeter film of oil
spread across the oceans of the world might have upon
the earth's refraction index. Should that index drop,
enormous climatic changes might ensue. To such scientists it seems foolhardy at best to proceed along the
path of high consumption and high pollution in hopes
that the cooling effects of such an oil film will somehow be offset.[12]

The third, and more limited line of ecological
argument is directed primarily at the specifics of oil
development in the Santa Barbara Channel. Advocates of
this line argue that the local environmental detriments

are not warranted because of the existence of at least
two greater alternative domestic oil sources. The first
of these is the extraordinary discovery of oil already
under development, on the north slope of Alaska. Reserves in that area are estimated at 40 billion barrels,
which is more than the estimated present reserves in
the entire continental United States.[13] Secondly, oil
shale deposits, found principally in the Rocky Mountain
region, are gigantic and constitute a potential oil
source for centuries to come. Studies by the United
States Geological Survey indicate that the shale reserves in sparsely populated areas of Colorado, Utah,
and Wyoming total around 4 trillion barrels of oil,
enough to sustain more than 360 years of production for
the entire world, based on the oil production rate in
1965.[14] Eighty billion barrels are recoverable under
present conditions and known technology; an additional
520 billion barrels of a relatively high grade oil
shale are potentially recoverable.[15] Although not
economically feasible at the present time, oil shale
resources should be considered in evaluating all future
development. However, large-scale production of oil
shale under present technological conditions may entail
enormous problems of waste disposal, water and air pollution, and severe damage to the landscape and vegeta-
tion.[16]

Who Benefits and Who Pays?

Broadly speaking, then, the analytical and political problems balance the national economic interest in
increased oil reserves and production against local
interest in an aesthetically pleasing community, and the
general public's interest in an ecologically viable
planet. However, even if one assumes with the oil industry that Channel oil production is economically sound,
despite its underlying subsidies, and that strategic
national needs override ecological spectres, there remains the problem of evaluating the national benefits
against the community's concern for an unimpaired seascape and clean water.

One approach to evaluation would require that all costs associated with oil development be charged to the companies engaged in development. These externalities of the companies' operations--pollution, beach and boat damage, loss to fishermen, etc.--would be internalized and would have to be calculated before undertaking development. This could be accomplished by putting a tax on "unintended flows" or pollution and by renegotiation of the "rentals"--the leases--to cover these costs. Unfortunately, virtually all of the tax and royalty benefits presently accrue to the national and state governments from offshore oil operations. To quote Santa Barbara Economics Professor Robert Weintraub: "Our problem is to persuade our parent governments to compensate us for at least part of the externalities we suffer from the Channel drilling operations."[17]

GEOLOGICAL KNOWLEDGE AND DRILLING PROCEDURES GOVERNING WELL A-21

There is no certainty that a blowout will not occur while drilling a land-based oil well, but lack of control on offshore wells presents far greater problems. A blowout on an offshore well can spread oil over thousands of square miles of water, with devastating pollution effects. Men, materials, and equipment are much more difficult to move to well sites for containment and cleanup procedures, the techniques of which are in themselves more difficult to execute offshore. Given these relative dangers, it is open to question whether current drilling technology will allow a sufficient margin of safety for developing offshore oil resources in areas of striking beauty. The geology, the blowout, and the attempts at oil containment in the Santa Barbara Channel provide an illuminating case in point.

The ocean floor of the Santa Barbara Channel forms a topographic and structural basin with an east-west axis. It contains numerous major and minor faults aligned with the structural trend. The general area is seismically quite active: Over 600 earthquakes have

been recorded in or near the Channel between 1934 and
1967.[18] Two severe quakes have occurred during the past
half-century--one in 1925 with a Richter magnitude of
6.3 and a second in 1941 with a magnitude of 5.9. The
stratigraphy of the Channel includes great layers of
Tertiary and Quaternary[19] sediments consisting primarily
of marine sandstones and shales, and one formation of
continental origin.

Typically, oil deposits in the Channel are apparently located in sandstones formed during the Oligocene,
Miocene, and Pliocene epochs;[20] however, detailed stratigraphy for each well is known only by the individual
oil companies on the basis of their own explorations
undertaken prior to the bidding for, and awarding of,
leases. At the time that leases are awarded, detailed
geological information is given to the Department of
the Interior, but it remains unavailable to the public.
According to one Assistant Secretary of the Interior,
by the summer of 1967 the oil industry had spent "a
couple of hundred million dollars"[21] in exploration
costs to gain information in the Santa Barbara Channel.
As the federal government has conducted no such program,
it must rely on the oil companies' information.

Whether the extent of the available information is
sufficient and its distribution is optimal from the
standpoint of anything but short-range corporate profits
are questions which remain beside known facts surrounding the oil spill. A more detailed review of those
facts than that offered in the introductory section of
this essay might raise not only doubts about whether
Union exercised sufficient prudence but also doubts
about whether, had the information about Channel drilling conditions been open to public scrutiny, the federal
government would have granted a variance from normal
drilling regulations.

The Blowout, and Efforts to Stop It

As previously noted, the blowout occurred while
Peter Bawden's drilling crew was removing the drill

string in order to run routine electric logs of conditions in the well hole.[22] Seven hundred and twenty
feet of drill pipe had been removed and stored in the
derrick on the platform when large quantities of drilling mud suddenly began shooting from the top of the
drill pipe section, which remained in the hole. Since
the top of this pipe was then some 90 feet above the
workers, the mud cascaded down over the floor of the
platform and made working conditions very difficult.
The mud flow soon subsided and was replaced by danger-
ously explosive gas. Lowering the top of the pipe to
the platform deck, the crew tried several unsuccessful
measures to bring the well under control. The gaseous
mist had completely engulfed the rig floor and visibil-
ity was severely reduced. Voice communications were
extremely difficult due to the roar of the escaping gas.
The drillers' eyes were affected by the toxic mist, and
an explosion seemed quite possible.

To make matters worse, one relatively easy way of
averting a threatening blowout was not available. In
the overwhelming majority of cases, a blowout threatens
first with a welling up of the drilling mud in the
annulus--the space between *the outside* of the drill pipe
and *the inside* of the well hole. Dealing with such a
problem generally entails only activating the blowout
prevention device which closes around the outside of
the hollow drill pipe and thus seals off the well.

In the case of Well A-21, however, the first observed sign of danger was mud *inside* the hollow pipe
shooting into the air. Once the mud had shot out,
nothing restrained gas or oil from coming up inside the
pipe. The best available evidence indicates that the
crew then tried a "second line" of defense--attempting
to screw a threaded blowout preventer into the top of
the drill pipe. But this procedure--not unlike trying
to replace the cork on a bottle of just-opened and
foaming champagne--also failed. Presumably, either the
threading on the preventer was damaged or the pressure
of the escaping gas was too great to screw it in. The
crew next sought to "stab" the top of the drill pipe

with the kelly--a long heavy shaft of steel used during
drilling to connect the drill pipe to the rotary table
which turns the drill. Apparently however, at some
point in these frantic efforts a break occurred in the
standpipe which carried drilling mud between the mud
pumps on the platform and the well hole. Thus stabbing
the hole with the kelly would have been futile even if
it had been possible: There would still have been no
way to pump mud back into the drill pipe to overcome
the oil and gas flow. Consequently all that could be
done was to close the blowout prevention device all the
way, which in turn entailed either pulling the pipe entirely out of the hole or dropping it back down to the
bottom. Since neither time nor circumstances permitted
the former, the pipe was dropped. The blowout preventer
was then completely closed and the blowout seemingly
contained. Unfortunately, fissures either preexisted
and intersected the hole below its casing or were created by a surge of pressure when the preventer was
closed. These fissures soon allowed gas and oil to
escape; hence, the large boils of gas and oil which soon
emerged on the ocean surface.

Could It Have Been Prevented?

In retrospect, the crucial question is whether or
not the blowout could have been prevented. One possibility is that the speed with which the drill string
was being removed was too great and that the effect
was to "swab in" the well. That is to say, *if* the
annulus just above the well-drilling bit becomes clogged
with cuttings of rock at some point during withdrawal
of the bit, and *if* withdrawal is not halted while drilling mud is pumped down to free the clogged annulus,
then the drill bit may function like a piston inside a
cylinder, the drill hole. It can then lower the pressure inside the hole--in effect creating a partial
vacuum. Thereupon oil in the surrounding oil-producing
horizon is likely to burst into the hole with great
force and push mud and gas up the drill pipe. That is
termed "swabbing in" a well. A number of Union critics

have--off the record--argued that this is the most likely immediate cause of the blowout.

Be that as it may, by Union's own admission it is likely that a blowout could have been prevented if enough casing had been utilized in the well.[23] By the time of the blowout, 13 3/8-inch conductor casing extended from the platform to a point 238 feet below the ocean floor. No other casing was set. Outer Continental Shelf Orders, promulgated by the Department of the Interior, require a minimum of 300 feet of conductor casing below the ocean floor and a minimum surface casing which totals 25 percent of well depths to 7,000 feet. Both of these requirements had been waived by the U.S. Geological Survey Regional Supervisor, D. W. Solanas.

Items appearing in the press shortly after the spill argued that California's offshore casing requirements were stricter than those of the federal government, and that had California standards been in effect the accident would never have transpired. However, examination of normal drilling practices in the Channel under both sets of regulations suggests that the situation is more complicated than that. In the Santa Barbara Channel, the initial drilling of an oil well is generally run down to at least 300 feet below the ocean floor without protective casing. California regulations then require the driller to undertake precautions against blowouts before proceeding below 350 feet, whereas the federal government permits as much as 500 feet of drilling before such precautions are undertaken. The precautions consist of installing casing from the floor of the drilling platform to the bottom of the partially drilled hole. Next, cement is injected around this first string of casing,[24] filling the gap between the casing and the hole made by the drill. Additionally, at least one blowout prevention device is attached to the top of the conductor casing at the platform. The drilling is then continued by means of a drilling bit of smaller diameter than that used before which runs down through the blowout prevention device and inside the conductor casing.

Under federal regulations the second stage of
drilling without further casing is permitted down to 25
percent of the proposed well depth or to 1,750 feet,
whichever is less. California regulations, on the other
hand, set a minimum of 1,000 feet and a maximum of 1,200
feet for halting the drilling once again, and depending
on well depth--may thus be looser or stricter. On
either showing, once the prescribed depth is reached,
drilling is again interrupted and a second protective
string of casing is placed down through the first string
of casing to the depth required by the standards of the
relevant jurisdiction. This string of casing is similarly cemented into place, and a vertical bank of several blowout prevention devices is attached to this
casing at the platform.

For four reasons heavy mud is continuously pumped
down during the drilling through the hollow drill string
to the drill bit and forced up to the surface again between the outside of the drill string and the inside
of the well hole. First, the mud carries the quarried-
out rock chips upward away from the working bit even as,
second, it cools the bit. Third, its weight helps to
prevent the escape of any high-pressure gases encountered in the subsurface strata. Finally, it prevents
the walls of the drill hole from collapsing.

After installation of the second casing string and
the second set of blowout prevention devices, the well
is normally drilled with--if necessary--similar stops
along the way to install more casing to the total depth
planned. When that depth is reached, a string of production casing is installed along the entire extent of
the well. This production casing is likewise cemented
to afford protection against high-pressure oil and gas
zones. Thus, hydrocarbons under pressure do not often
escape from cased wells. During drilling, the mud
column in the hole normally prevents such an occurrence.
Should the column of mud prove too light to restrain
the oil or gas, the blowout prevention devices should
still be effective. However, if faults or fractures in
the earth's strata slant through a partially cased

drill hole up toward the surface of the ocean floor,
then, quite clearly, oil and gas are likely to appear
on the ocean's surface. The same result can occur if
the sands or rock surrounding the hole just below the
lowest part of the casing are too porous or weak to
withstand a sudden increase in pressure produced by oil
rushing up from greater depths. One or both of these
phenomena evidently occurred in the Santa Barbara case.

The Fatal Flaw

The foregoing description strongly suggests that
the fatal flaw in the A-21 drilling practices lay, as
far as the federal government was concerned, not in the
normal regulations but in granting the variance. There,
perhaps, is the fundamental relevant difference: California statutory standards appear to be more strictly
adhered to.[25] On April 12, 1969, as part of the post
mortem, Frank Hortig, Executive Officer of the California State Lands Commission, was asked if California had
ever granted a variance permitting the elimination of a
second string of casing on any offshore well in California State waters. He replied, "The answer is an un-
qualified no!"[26]

After the spill, United States Geological Survey
spokesmen insisted that sound engineering practices
dictated the casing variance on Union's well.[27] However, these sound engineering practices have not yet
been disclosed to the public in sufficient detail by
either the oil companies or the Department of the
Interior. Until they are, there is little reason for
the public to be convinced of the propriety of allowing
a variance on the second string of casing, a device
which might have prevented massive pollution of the
Channel. Since the spill, moreover, final authority to
grant variances for the elimination of a second string
of casing for Channel wells has shifted from the regional headquarters of the USGS to Washington.[28] It
seems unlikely that this authority will be overworked.

REMEDIAL MEASURES

Shortly after the blowout, Union Oil Company commenced massive attempts to kill the well. A team from
the Red Adair Corporation was called in a few hours
after the accident, followed soon after by a group of
geological experts from the Department of the Interior.
Millions of pounds of mud and cement were used in attempts to close the well in, but all efforts succeeded
only in diminishing the flow, not in stopping it entirely.

Similarly, Union Oil's attempts to restrict the
damage done by oil already spilled were also unsuccessful. By February 12, 1969 the company had 37 supervisors directing the efforts of nearly 800 workmen in
containment and cleanup. Union also marshalled 18,900
feet of booms designed to restrict the spread of oil,
122 trucks, bulldozers, tractors, and other pieces of
heavy equipment for cleanup work, and 54 boats and crop-
dusting planes to spray chemical dispersants.[29]

Dispersants of numerous sorts--including Guardian
Chemical's Polycomplex A-11, Esso's Corexit, and Ara
Chem--were utilized in the attempt to speed bacterial
breakdown of the oil. Yet anxieties resulting from the
Torrey Canyon incident about dispersant toxicity became
a matter of dispute between conservationists and manu-
facturers, and the Coast Guard severely limited their
application. Heavy seas, inadequate methods of application, and high cost further restricted their use and
effectiveness.

Mechanical devices, chiefly floating booms, were
tried with little more success. A hollow plastic boom
hastily installed across the entrance of Santa Barbara
Harbor to protect moored yachts was punctured by the
prow of an errant boat and collapsed. Heavy ocean
swells first delayed and then virtually rendered useless
a V-shaped boom towed near the platform by two tugs.
After two days of unsuccessful efforts by a barge situated at the apex of the V to suck up the oil, this boom

was retired to port for repairs. A second skimming device--a boat equipped with a 21-foot wide suction nozzle and a tank for separation of oil and water--had but a modest success: Its recovery rate was limited to a few barrels per hour. Also modest in its recovery rate was a large iron funnel placed over the source of the spill. During the summer of 1969, Union experimented with placing 115-foot-square rubber mats on top of the leaking ocean fissures. By February 1970, a year after the spill, the best device for catching the oil at its source appeared to be Union's installation of funnel-shaped plastic tents directly above the fissures on the ocean floor. According to the perhaps optimistic estimates of the USGS, these funnel-shaped tents were catching about 25 percent of the escaping oil.[30]

In respect to cleaning the beaches, American ingenuity seemed sorely, almost humorously, taxed by nature here. High tides covered the rocks with oil once again after steam cleaning and sandblasting, while placing clean sand over the oil-soaked beach depended for its success upon the absence of shifts in prevailing swells. Such a shift would probably disclose the mess beneath. Only physical removal of the oil-soaked sand and the scattering and collecting of oil-absorbent straw appears as yet to have been effective.

Since the spill, the oil companies and governmental agencies have received over 3,000 suggestions from the private sector. Thus far, the most promising appear to be those which focus upon gathering spilt oil near its source before it has had time to spread. Thus Garrett Corporation, an aerospace firm located in Los Angeles, has received a $600,000 contract financed by the Federal Water Pollution Control Administration and the American Petroleum Institute to develop its Sea Dragon prototype. The Sea Dragon is a boat containing a large centrifugal separator which operates at 4,000 r.p.m. According to company claims, it will be able to sweep and skim 128 acres of ocean surface per hour while separating oil from water at the rate of 25,000 gallons per hour, returning the water to the ocean and storing the gathered oil. The prototype model underwent tests during the summer of 1970 but has not been widely adopted since that time. Standard Oil of California has tested in the Channel another containment device, consisting of floating drums, wood panels, and plastic sheets, which it hopes will constitute proof of the oil industry's ability to contain any future blowouts. Standard hopes this will convince the reluctant California State Lands Commission to issue it a new permit to drill for oil in state-owned waters.[31]

It is possible that one or both of these devices will prove that the industry has developed adequate remedial technology. However, to many Santa Barbara citizens the most plausible moral to be drawn from the "cleanup story" is the fact of insufficient pretesting of cures by industry and government. Although Santa Barbarans are no doubt biased on this score, in June 1970, American insurance firms virtually confirmed their suspicions by refusing from that date forth to cover the oil companies against liability for spills emanating from offshore platforms.[32] As stated by the Insurance Rating Board, theretofore coverage had not included release of effluents to the air, water or land, "except when such discharge is sudden and accidental."[33] Thereafter, pollution of water by oil was explicitly excluded, whether or not sudden or accidental. The board stated:

> This exception for oil is justified
> because oil spillage into water is a
> catastrophic phenomenon of recent oc-
> currence and not contemplated when
> existing rates were made. The public
> has become increasingly aware of agon-
> izing effects of oil spillage and re-
> sentful of the nature of the failure
> of oil risks to take the ordinary
> steps to avoid them.
>
> To ignore the attitude of society
> toward the destructive effects of oil
> spillage by making insurance readily

> available, as part of underlying
> coverage, without regard for the
> exposure, could be interpreted as
> a disregard for the public interest
> and the public policy.
>
> Because of this situation, any com-
> panies that may afford this insur-
> ance will undoubtedly do so only
> after a careful survey of the expo-
> sure and the methods for preventing
> oil spillage into water in order to
> assure protection for the public
> interest.[34]

The oil companies themselves have acted to provide insurance coverage until such time as regular carriers might make such a "careful survey", by creating their own insurance company. This company is appropriately called Oil Insurance Limited (OIL) and is the creation of several major companies such as Atlantic, Richfield, Cities Services, Gulf, Signal, Standard of California, Marathon, Phillips and Union. The purpose of this company is to protect the companies against catastrophic events involving "on-shore and off-shore property, pollution and bringing under control wild oil or gas wells or extinguishing oil or gas well fires."[35] Each stockholding company is required to subscribe $10,000 per share of capital stock, for which it receives coverage up to a maximum of $100 million (minus a deductible ranging from one to ten million dollars) in any one year or in respect of any one occurrence. While the withdrawal of insurance coverage for oil spills created problems for the oil industry, it does not appear that it will constitute a serious impediment to further off-shore exploration and development.

CHAPTER V

Administering Offshore Oil:
A Study in Divided Responsibility

"COOPERATIVE FEDERALISM" WITHOUT COOPERATION

The essence of a well-functioning federal system is found in government's capacity to discriminate and allocate resources between those public concerns best managed, regulated, and promoted at the national level, and those best dealt with at subnational levels.

Originally conceived of as a "layer-cake" arrangement, with local governments at the base, states in the middle, and the federal government at the top--and with each successive layer further removed from the people-- it is now generally agreed that the "marble-cake" imagery is more descriptive of contemporary American federalism. Now there is "an inseparable mingling of differently colored ingredients, the colors appearing in vertical and diagonal strands and unexpected whorls. The colors are mixed in a marble cake, similarly, functions are mixed in the American federal system."[1] The supposed result is "cooperative federalism," involving the federal government in complex financial, administrative, and program relationships with both state and local governments.[2]

Early commentators such as James Madison viewed the national government as a more likely defender than local government of the general interest because of the great multiplicity of factions found in a nationwide representative body which no one faction could dominate.[3] Modern exponents of this position conclude that states tend

to be systems "of protection for favored groups."[4]
Through fractionation of power and informal ties it is
alleged the states have become servants of special interests rather than of the general welfare. Not long
ago, one critic of the federal system concluded that
federalism has worked chiefly to the advantage of capi-
talists, landlords, and racists.[5]

The politics surrounding the Santa Barbara oil
spill raise questions about the conventional wisdom concerning the manner in which the contemporary federal
system allocates scarce resources. While heavily criticized both as instruments for the protection of special
economic interests and as a means of permitting the op-
pression of ethnic minorities, might not states, counties, and cities also protect certain public interests
that could be adversely affected if left solely to private economy and the national government? Like local
interests fighting federally sponsored urban renewal,
freeway construction, and automobile pollution, might
not the people and local government agencies of Santa
Barbara be struggling to protect a precious long-run
value--an unpolluted and aesthetically pleasing environ-
ment--against a much shorter term interest in federal
revenues and oil company profits?

The management of oil-bearing submerged lands and
the control over the waters above those lands highlights
the jurisdictional quagmire and competition between mul-
tiple agencies with restricted responsibilities. Pollution resulting from actions taken by an operator subject
to the jurisdiction of one agency affects the programs
and responsibilities of other agencies. The same water
provides for navigation, recreation, fishing, and visual
appreciation, which are not usually complementary, or
necessarily compatible with oil production. Regulatory
conflict is a natural result.

In at least some respects this jurisdictional complexity creates undesirable competition. Since passage
of the Submerged Lands Act of 1953, jurisdiction over
submerged land has been divided between the federal and

state governments, with the boundary at the three-mile
limit. The oil-bearing strata, however, constitute a
common pool, and withdrawals from the seafloor in one
jurisdiction tend to deplete the entire pool. Thus,
oil production on state leases in the Summerland area
east of the Santa Barbara State Sanctuary impinges upon
resources under the federal jurisdiction and encourages
federal oil development. Similarly, granting leases in
federally controlled areas threatens the existence of
the State Sanctuary. The jurisdictional tangle is further complicated by the fact that optimal resource extraction requires management to ensure economical and
efficient production. Too rapid extraction leaves much
oil in the ground, beyond recovery. The possibility of
conflicting approaches to recovery is obvious, and the
consequences of efficient development may be serious.

Even if federal and local interests were always
complementary, governmental stratification, both vertical and horizontal, would still tend to complicate management. With authority as widely dispersed as it was
during the late 1960s, it was almost inevitable that
ample responsibility would be felt by no single agency.
At the time of the spill many governmental entities
had a piece of the action, but none took an overall approach to the ecology of the area.

FEDERAL ADMINISTRATION OF OFFSHORE OIL DEVELOPMENT

Under the provisions of the Outer Continental Shelf
Lands Act (OCSL) passed in 1953,[6] the Secretary of the
Interior was delegated the authority to lease submerged
lands beyond the three-mile limit on the Continental
Shelf.[7] Leasing programs have been administered by the
Bureau of Land Management, relying to a considerable
extent on data supplied by USGS. The Secretary is authorized to prescribe such regulations as he might deem
necessary in granting such leases. Supervision of operations conducted subsequent to the granting of leases
is the responsibility of USGS. Its regulations include
rules "to prevent damage or any waste of any natural

resource, or injury to life or property."[8] The regulations cover drilling, well spacing, well casing, approval of plans for discovery and development, keeping of
records, samples, and surveys, and control of wells.
The OCSL Act authorized the Secretary of the Interior
to require the prevention of pollution in offshore oil
or mining operations. The Code of Federal Regulations
also provide that:

> The lessee shall not pollute the
> waters of the high seas or damage
> the aquatic life of the sea or al-
> low extraneous matter to enter and
> damage any mineral or water-bearing
> formation. The lessee shall dis-
> pose of all useless liquid pro-
> ducts of wells in a manner accept-
> able to the supervisor.[9]

At the time of the spill, Department of the Interior also administered the recently repealed Oil Pollution Act[10] of 1924, which, as aménded in 1966, prohibited "grossly negligent" or "willful" discharge of oil
into navigable and shoreline waters from a vessel. The
1924 Act established criminal sanctions but provided
some exceptions in the cases of emergencies, unavoidable
accidents, collisions, and certain permitted discharges.
In cases of pollution endangering the health and welfare
of persons in the same state where the pollution discharges occurred, the Secretary of the Interior was
authorized to request the Attorney General of the United
States to bring suit for abatement of the pollution
with the consent of the governor of that state.

The Federal Water Pollution Control Administration--
recently renamed the Federal Water Quality Administration[11]--has an interest in spills such as that in the
Santa Barbara Channel since it was created as a branch
of the Interior Department for the purpose of cooperat-
ing with the states in the improvement of water quality
in interstate waters.[12] The FWPCA has worked chiefly
through state and local administrative units in

developing comprehensive pollution control programs.
The role of the FWPCA in the Santa Barbara oil spill
was chiefly technical; providing sanitary engineers,
chemists, biologists, and oil pollution specialists,
and exercising influence over Union's use of dispersants.[13]

The Army Corps of Engineers has the authority to
issue licenses for structures in or over the navigable
waters of the United States. The OCSL Act extended
this authority to artificial islands and fixed structures on the Outer Continental Shelf in the interest of
protecting navigation.[14] The Corps of Engineers also
issues permits for geophysical and seismic explorations
in the Channel; core drilling permits must be obtained
from the corps. The Refuse Act of 1899[15] is likewise
administered by the corps and applies to virtually all
discharges into navigable waters from either vessels or
shore facilities.

However, not considering the prospect of pollution
to be a matter of its concern in issuing permits, the
corps leaves this responsibility to the Department of
the Interior. When the application for the first platform on the Outer Continental Shelf in the Santa Barbara Channel was being considered by the district engineer in 1967, a question was raised whether the corps
could properly consider objections based on the possibility of pollution. The district engineer testified
in 1969, after the spill, that Interior and Corps of
Engineers officials had agreed:

> all aspects of the public interest
> other than navigation and national
> security will be considered and
> evaluated by the leasing agency
> prior to granting oil drilling
> rights...the Department of the
> Interior should, therefore, prop-
> erly exercise primary cognizance
> of items affecting fish and wild-
> life, pollution, and at least some

other aspects of the ecology and
the aesthetics of the area.[16]

Finally, the United States Coast Guard, functioning
as a service under the Department of Transportation, is
a major participant in control and cleanup of pollution
on coastal waterways. The Coast Guard imposes controls
over the discharge of pollutants from platforms, pipelines, and tankers. Under the OCSL Act the Coast Guard
is responsible for administration of safety provisions
having to do with navigational hazards and prevention
of collisions. It maintains surveillance, along with
the Corps of Engineers, over coastal waters for detec-
tion of pollution.

Under the National Multi-Agency Oil and Hazardous
Materials Contingency Plan approved by President Johnson
in 1968, the Coast Guard was given a major responsibility in pollution control. The purpose of the plan was
to provide a coordinated response to pollution inci-
dents. Its four major objectives were:

1. to develop effective systems for discovering
    and reporting the existence of a pollution
    incident;

2. to promptly institute measures restricting the
    further spread of pollutants;

3. to apply techniques to clean up and dispose of
    the collected pollutants;

4. to institute action to recover cleanup costs
    and effective enforcement of existing Federal
    statutes.[17]

The plan specified that On-Scene Commanders take charge
whenever a pollution incident occurred. In the Santa
Barbara incident, the Coast Guard lieutenant in charge
of the Santa Barbara Coast Guard group acted as the On-
Scene Commander, coordinating all efforts to contain the
spill and coordinating and approving all methods aimed
at cleaning up the oil.

STATE ADMINISTRATION OF OFFSHORE OIL DEVELOPMENT

Responsibility for administration of submerged lands under the jurisdiction of the State of California is given to the State Lands Commission, consisting of the Lieutenant Governor, the State Controller, and the State Director of Finance. The commission is authorized to grant leases when they are in the best interests of the state.[18] California statutes provide that:

> Pollution and contamination of the
> ocean, and tidelands, and all impair-
> ment of and interference with bathing,
> fishing, or navigation in the waters
> of the ocean or any bay or inlet
> thereof is prohibited, and no oil,
> tar, residuary product of oil or any
> refuse of any kind from any well or
> works shall be permitted to be de-
> posited on or pass into the waters
> of the ocean or bay or inlet
> thereof...[19]

The commission is thus charged to determine whether proposed leases would:

> Be detrimental to the health, safety,
> comfort, convenience, or welfare of
> persons residing in, owning real prop-
> erty, or working in the neighborhood
> of such areas; interfere with the
> developed shore line, residential or
> recreational areas to an extent that
> would render such areas unfit for
> recreational or residential uses or
> unfit for park purposes; destroy, im-
> pair, or interfere with esthetic and
> scenic value of such recreational,
> residential or park areas; create
> any fire hazard or hazards, or smoke,
> smog or dust nuisance, or pollution
> of waters surrounding or adjoining
> said areas.[20]

The State of California has an obvious interest in preventing pollution of its adjacent waters and at the time of the spill already had three major statutory enactments dealing with pollution. Unlawful pollution was defined by one law as "impairment of the quality of the waters of the State by sewage or industrial waste," which "adversely and unreasonably affect the ocean waters and bays of the State devoted to public recreation."[21] Administration of this law was given to state and regional water pollution control boards. These boards are empowered to investigate conditions of pollution, to issue waste discharge requirements, and, after hearings, to issue cease-and-desist orders. Failure to comply can result in a request for an injunction and ultimately in civil monetary penalties of up to $6,000 per day.

The Fish and Game Department of California enforces a provision of the Fish and Game Code which makes it unlawful to discharge into the waters of the state "any petroleum" and numerous other substances.[22] Violation of this statute calls for the imposition of criminal penalties. Finally, the Harbors and Navigation Code imposes a civil penalty of up to $6,000 plus the cost of cleanup and damages for allowing oil to be dispersed in the waters of the state, whether intentionally or negligently.[23]

Through both formal and informal relationships, federal and state agencies managed to achieve a reasonable level of cooperation under the crisis conditions occasioned by the oil spill. Indeed, Mr. Paul De Falco, Jr., Regional Director of the FWPCA, observed:

> State and federal agency personnel
> have worked together in such a fashion
> that it has been difficult at times to
> tell who belonged to which agency.
> There has been a recognition that the
> incident requires a response to the
> best of the abilities of all of the
> individuals concerned without regard
> to parent agency loyalties.[24]

But, while goodwill and a common professional outlook
may unify officials at the operating level, particularly
amidst a crisis, neither can substitute for overall co-
herence of policymaking. Because federal and state
agencies sometimes have different priorities, it is not
surprising that there should be conflicts over offshore
oil development. During the 1960s this altogether too
splendid array of agencies, administering diverse laws,
was almost bound to result in ineffective regulation
and inadequate protection of the not-so-divisible
Channel ecosystem.

.

CHAPTER **VI**

Executive and Legislative Remedies:
Federal and State

To many Santa Barbarans, the course of events leading up to and following the January 28 spill constituted an excellent, if unpleasant, illustration of what was wrong with the Topsy-like administration of offshore oil development in general. Not surprisingly, the overwhelming hope of the local public was for a quick response by the federal government, termination of all Channel drilling in federal waters, and restoration of Santa Barbara's aesthetic and ecological *status quo ante*. However, given the obvious disparities between the political clout of a small city and that of a gigantic international industry, plus the inherent difficulties of unravelling such an administrative knot, it was not at all obvious, as of February 1969, that any single arm of the federal government possessed the authority to produce such a solution. Nor, given the ecological damage done and the state of remedial technology, was it obvious that *anybody* had the power or ability to restore past conditions. An analysis of the remedies available--both those effected by the national government and those being pursued by Santa Barbarans must consider the constraints of both positive and natural law.

The primary constraint of positive law is, of course, that mandated by the United State Constitution respecting the vested rights deriving from the terms of Channel leases and from the oil industry investments in developing the leases. Whatever the future constitutional status of the various claims and claimants,

executive and legislative remedies must respect the
industry's solid constitutional position.[1]

The question is which of the various constitutional
rights claimed by various companies since the blowout
require governmental deference? Possibilities range
from one that would allow very little redress of Santa
Barbaran grievances to one that would permit a close
legislative approximation of the local public's maximal
reasonable expectations. At one hypothetical end stands
the assertion that the federal government, by the act
of accepting a lease-bid in February 1968, may not
change the drilling regulations and exploration rights
then in force because there is no just means whereby
the federal government can calculate and compensate for
changes in industrial profits deriving from changes in
the original contractual terms of the lease. While no
oil company has quite maintained this, the action of
Pauley Oil Company[2] and a number of other minor producers filed in federal court in March 1969 tends to
follow this reasoning. Pauley Petroleum argues that
any changes in drilling regulations--be they temporary
drilling halts called for by the Secretary of the
Interior or stiffer drilling regulations imposed by him--
which make development more expensive than anticipated
at the time of bidding, thus reducing the industry's
anticipated profits, must be compensated by the federal
government. At the other hypothetical end is the position that the federal government may do what it wishes,
even terminating the leases, as long as it provides acceptable compensation either in monetary terms or by
arranging swaps for other oil-bearing federal lands
located elsewhere.

INITIAL FEDERAL REMEDIES

The federal government can undertake one or more
of seven types of possible remedy, depending upon the
extent to which it feels constrained by constitutional
questions and the financial burdens to the taxpayer.
Three of these bear primarily upon the specifics of the

Santa Barbara Channel situation and, arguably, could be
achieved by executive authority alone. Although all
three could be given supportive congressional legislation, thus cloaking them with the illusion of perman-
ence, they do not seem to require legislative approval.
These first three remedies consist of: (1) strengthen-
ing the preventive, safety requirement of drilling
regulations; (2) advancing the technology of "cure";
and (3) limiting the areas in the Channel subject to
drilling.

Stronger Drilling Regulations

Drilling regulations in effect when the blowout
occurred were established by the Department of the
Interior and were supposedly designed to carry out legislative intentions to safeguard the public interest.
As long as regulations remained consonant with such intentions, they could be modified with due notice. As
noted earlier, one of the most persistent criticisms
leveled at the federal government after the spill was
that the Department of the Interior's regulations, and
the USGS's administration of these regulations, were
less stringent than the rules and practices followed by
the State of California.[3]

One of the earliest reactions of the federal government was to tighten drilling practices in federal
waters off California. The new regulations, promulgated
in May 1969, as OCS Order No. 10, greatly broadened the
authority of the Department of the Interior to circum-
scribe the behavior of the lessees and impose significant new obligations on them. A supervisor in the Department of the Interior is authorized to suspend any
operation which in his judgment "threatens immediate,
serious, or irreparable harm or damage to life, includ-
ing aquatic life, to property, to the leased deposits,
to other valuable mineral deposits or to the environment."[4] Earlier regulations simply required the lessee
to "take all reasonable precautions for keeping oil
wells under control at all times."[5] The new regulations
stated more broadly that the lessee "shall not pollute

land or water or damage the aquatic life of the sea...."[6]
And he shall also "keep oil wells under control at all
times...."[7] Thus, there appeared to be no "escape
hatch." The newer regulations were much more detailed
than the older ones in providing for casing and cement-
ing of wells, the use of drilling mud, and the provision
of blowout prevention devices. Lastly, the regulations
required the lessees to provide much more detailed information about their development plans. Writing to
Senator Henry M. Jackson, Chairman of the Senate Committee on Interior and Insular Affairs, on May 16, 1969,
Russell E. Train, then Undersecretary of the Interior,
stated that the Department had "put into effect the most
stringent possible regulations we can devise to safeguard our environment."[8]

Technological Limitations

From the standpoint of the Santa Barbara citizens,
the trouble with these new regulations was twofold.
First, as they were preventive in nature, they did not
cure the continuing spill. Second, they did not guar-
antee that another spill would not be forthcoming, or
directly speak to the major local objection that, even
without spills, the rigs themselves were a monstrous
aesthetic affront to Santa Barbara's scenery and a de-
triment to its capacity to attract tourists and wealthy
retired persons.

During summer 1969 the federal government developed
a plan of remedy for the spill, which purportedly attempted to meet the first objection. In June 1969, a
high level commission appointed by Dr. Lee DuBridge,
President Nixon's scientific adviser, recommended a pro-
cedure that delighted Union Oil Company but affronted
many citizens of Santa Barbara. After hearing closed
testimony from oil company experts and officials of the
USGS, the DuBridge commission recommended an acceler-
ated program of well drilling from Platform A in order
to reduce the underground pressure that was forcing the
oil to the surface of the ocean floor.[9] Santa Barbarans

were not pleased when told this remedy would probably
work in 10 to 20 years' time. It is with respect to
the technology of cure that the federal government and
the oil industry remain most open to criticism.

Limiting Areas of Drilling

At the same time, the Department of the Interior
ventured into a third type of remedial action--limiting
the area of the Santa Barbara Channel subject to drilling. It converted the existing two-mile federal buffer
seaward of the Santa Barbara State Oil Sanctuary into a
permanent ecological preserve of 21,000 acres. In addition the Department of the Interior created a new federal buffer zone of 34,000 acres south of the ecological
preserve, where neither drilling nor production would
be permitted. Finally, the department promised not to
open any unleased federal Channel lands for bidding
without first consulting the public.[10]

THE DUBRIDGE REPORT AND LOCAL DISSATISFACTION

Early in the summer of 1969, the Department of the
Interior argued that these three types of changes in
the ground rules for drilling in the Santa Barbara Channel constituted a proper balancing of local Santa Barbara interests and the broader public interest in development of the nation's oil resources. In the government's view, not surprisingly, those changes within the
executive branch were sufficient. But the reaction of
the contending sides, however, indicated that the compromise was not acceptable to all parties. Senator
Cranston, Congressman Teague, city and county officials,
the Santa Barbara *News-Press*, and local conservationist
groups denounced the DuBridge Commission recommendation
to no avail.[11] The commission claimed that abandoning
the leaking well, or attempting further to seal it off,
might prove more dangerous than speeding up the pace of
oil withdrawal. However, the refusal of the commission
to make public either its hearings or the details of

its report simply fortified local animosities and suspicions against the executive branch. Soon the ACLU
(American Civil Liberties Union) joined the city, the
county, and 17 private individuals led by former Democratic State Senator Alvin Weingand, in bringing suit
to force public disclosure of the evidence and reasoning of the DuBridge Commission.

Union Oil lost no time in taking the commission's
advice. By the end of April 1970, Union President Fred
Hartley was able to report to his company stockholders
that a second platform situated near Platform A had
been completed; that the two together were extracting
about 45,000 barrels per day from 42 wells; and that in
consequence the Channel had become the fifth largest
oil-producing field in the California area. So successful, indeed, was the pressure-reducing program advo-
cated by the DuBridge Commission that Hartley anticipated that the Channel would soon become the fourth
largest field.[12]

While Santa Barbara residents were not unhappy with
the tightening of drilling regulations, they perceived
the administration's buffer zones as unsatisfactory.
There were two chief reasons for this reaction. First,
the 55,000 acres withdrawn from leasing by Secretary
Hickel was barely one-tenth the size of the Santa Barbara Channel acreage opened up for bidding at the time
of the first leasing program in late 1967. Of the
540,600 acres split into 110 tracts of roughly 9 square
miles each, only 71 tracts had received bids sufficiently high to be financially acceptable to the government. Many would not receive substantial bids even if
the government reopened them for leasing in the future.
But the principal local objection was that the administrative action by its very nature appeared far from
final. What could so easily be withdrawn in 1969
could, presumably, be as easily opened up again at a
later date. And while the local community was pleased
to find two companies turning back leases voluntarily
to the government as they proved unfruitful,[13] there
remained a large amount of leased territory where the

appearance of platforms seemed to detract seriously from the beauty of the Channel.

CITIZEN ACTION

Local citizen groups interpreted administrative actions as distinctly partial toward the oil industry, and increasingly they turned to self-help to dramatize their position. One tactic was picketing, led by GOO, of the wharf in the Santa Barbara Harbor used as the shore base for oil company supplies. The high point came in November 1969 when, frustrated in their attempts to persuade the Army Corps of Engineers to hold a public hearing before issuing a drilling permit to Sun Oil Company, GOO determined to test the rights of fishermen against those of the oil industry. Shortly before Sun Oil's Platform Hillhouse reached its destination via barge, a GOO yacht sailed from Santa Barbara Harbor to the presumed location Sun Oil intended for the platform, dropped anchor, and began to fish. GOO proposed to fish there until the boat's timbers rotted away or until the Coast Guard should interfere-- whichever occurred first. On November 23, GOO appeared to have frustrated the Sun Oil Company. Great was the local rejoicing when the oil crew on the barge miscalculated and the entire platform fell over into the Channel. Three days later however, GOO's brief victory evaporated. Retrieving the platform, the barge crew set it down a few hundred feet away from the anchored GOO yacht. GOO, it seemed, had slightly miscalculated Sunoco's logistics.[14]

PROPOSED NEW FEDERAL LEGISLATION

While local interests were being graphically articulated on the home front, pressure was being exerted in Washington for a more "permanent aggregation of interests" in the form of four types of remedial federal legislation: (1) changing the general tax treatment of oil to make high-risk drilling less attractive;

(2) changing the terms of liability so that the oil industry might exercise more caution; (3) making it easier for citizens to bring charges against polluters in general; and (4) phasing out as many Channel leases as possible.

Tax Reform and Industry Liability

As late as early 1972, only the first two types of legislative remedies appeared fruitful. After a lengthy battle, the tax reform legislation passed in 1969 had reduced the oil depletion tax allowance by 5 percent. Further, both a presidential task force and a majority of the cabinet had recommended overhauling the oil import quota system in favor of tariffs.[15] However, it should be noted that the President himself had undertaken no action to promote this change, and the troubled Middle Eastern situation had, by July 1970, temporarily undermined the chief argument against the quota system-- that it forced the American public to pay artifically high prices for domestic oil. On May 3, 1970, the Trans Arabian Pipeline had been ruptured by a bulldozer, and the Syrian government refused to permit its repair, blocking the flow of 500,000 barrels per day of Saudi Arabian crude oil from reaching the eastern Mediterranean. At the same time the Libyan government reduced oil production by the same amount. The loss of this oil forced western importers to resort to tankers. That, in turn, forced the spot charter price of tankers up to record levels, 50 percent higher than during the 1967 Arab-Israeli War, doubling the previous year's price of Persian Gulf oil coming into the United States. At the time of the Santa Barbara oil spill domestic oil was about $1.00 more expensive than Middle Eastern oil delivered to East Coast refineries; in summer 1970 the gap had closed and a reverse differential of about $.75 per barrel developed.[16] While the short-term domestic effect of this price change was almost sure to take some of the political punch out of arguments decrying the artificiality of U.S. import quotas, the phenomenon of relatively speaking "bargain" domestic U.S. oil was

temporary. By the end of 1971 the Middle Eastern oil
was once again flowing through TAP-line and domestic
U.S. oil was again more expensive than world oil delivered to the East Coast.

Water Pollution Legislation

Progress was made along a remedial path on April 3,
1970, when the President signed P.L. 91-224, the Water
Quality Improvement Act of 1970.[17] The act repealed
in its entirety the 1924 Oil Pollution Act and its 1966
amendment, thus accomplishing what Congressman Charles
Teague had sought earlier by introducing into the First
Session of the 91st Congress a bill to eliminate the
1966 provision of limited liability to grossly negligent
or willful oil spillage.[18] The 1970 act, beginning with
expansive definitions of "Oil," "Discharge," "Vessel,"
"On Shore and Off Shore Facilities," made unlawful any
discharge of oil into United States waters and held the
owner or operator of a vessel or facility liable unless
he could "prove that a discharge was caused solely by
(A) and act of God, (B) an act of war, (C) negligence
on the part of the United States government, or (D) an
act or omission of the third party...."[19] More impor-
tantly, it shifted to the owner or operator of such a
facility the onus of demonstrating that such a discharge
fell within any of the enumerated sections.

In the event of inability so to prove, liability
to the United States government for cleanup costs could
run, in the case of onshore and offshore facilities, as
high as $8 million and, in the case of vessels, as high
as "$100 per gross ton of such Vessel or $14 million,
whichever is the lesser."[20] Further, if the federal
government could prove that the discharge resulted from
willful negligence or willful misconduct on the part of
the owner or operator, there then would be unlimited
liability for full removal costs. The act obliged the
oil industry to report spills by imposing a civil penalty of up to $10,000 for each failure to report.[21]

Beyond that, Congress stated that the act was in
no way meant to abrogate any other liability to private
persons which might accrue from oil damages; and it expressly permitted the state and local governments to
impose further requirements and liabilities for discharging oil into state-owned waters.[22] While this act
did not broaden the liability of the Union Oil Company
spill in the Channel, it did indicate a substantial
shift in congressional attitudes towards industrial
polluters.

Facilitating Court Challenges

During the 91st Congress (1969-70) the Senate also
considered bills which would redress the balance of
power in two further ways. First, and like the Water
Quality Improvement Act, which was directed to the general issue of offshore pollution rather than to the
specifics of the Santa Barbara Channel controversy,
S.3575, introduced by Senator George McGovern of North
Dakota, would permit any individual to bring an individual or class suit for declaratory or equitable relief
against any polluter of air, water, or land. Finding
constitutional support in the Commerce Clause, McGovern's
bill would require only that the plaintiff procure supporting affidavits from "two technically qualified persons stating that to the best of their knowledge the
activity which is the subject of the action damages or
reasonably may damage the air, water, land, or public
trust of the United States by pollution, impairment or
destruction."[23]

Prohibition of Drilling and Lease Termination

A second type of proposed bill dealt specifically
with the issue of drilling and development in the
Channel itself. During the First Session of the 91st
Congress in 1969 California Senators Cranston and Murphy
and Congressman Teague had introduced, in their respective Houses, bills which in varying measure sought to
prevent further exploitation of the Channel's oil

resources.[24] The Nixon administration had opposed these bills as unnecessary in May 1969; however, by late spring 1970, the administration had reversed itself. On June 24, 1970, Senator Murphy, acting for the executive branch, introduced a bill into the Senate which would have rescinded 20 of the 71 leases and provided a mechanism for compensating the oil companies.[25]

Congressman Teague introduced legislation in 1969[26] and again in 1970[27] which would rely on the Naval Petroleum Reserve of Central California--the so-called Elk Hills Reserve--as a means of compensating present lessees for their rights in the Santa Barbara Channel. The first bill would have created an oil reserve in the Santa Barbara Channel, cancelled all leases in that reserve, and would have given lessees the option of a cash settlement or rights in the Elk Hills Reserve. The State of California would have had to agree to suspend all operations in the tidelands to protect the federal interest in the offshore reserve.

The second bill would have created a national energy reserve and terminated all leases in the Channel. The act would have created a Santa Barbara account, the funds for which would be derived from sales of oil produced in the Elk Hills Reserve. These funds would be used to compensate lessees who were authorized to bring action against the United States asking compensation for the cancelled leases.

While matters proceeded slowly in the House of Representatives, on July 23 and 24, the Senate Interior Subcommittee on Minerals, Materials, and Fuels, chaired by Democratic Senator Frank Moss of Utah, heard testimony on all the Channel bills. Senator Cranston argued strongly along with Santa Barbara citizens and local governments, that the broadest bill, S.1219 should be reported out by the subcommittee. This bill would have terminated, with compensation, all the leases in the Channel (except two that were operating to reduce pressure, according to the DuBridge Commission recommendation, and a third that was taking oil from the same

field in state jurisdiction east of the sanctuary).
Then Secretary of the Interior Hickel argued for the
administration's more modest bill. From the start, it
was not at all clear that any of the bills would receive
the blessing of the subcommittee majority. However, it
was intriguing to find the administration switching to
a position that was opposed by both the oil companies
(on the grounds that withdrawing 20 leases would unnec-
essarily impede development of the nation's resources),
and also opposed by Santa Barbara citizens (on the
grounds that it did not withdraw enough).[28] Nonetheless, the administration's position could be considered
a potentially significant harbinger for two reasons.
First, S.4017 made plausible Secretary Hickel's October
1969 statement that there would be no more Channel
leases for sale.[29] Second, it indicated that a possible
wholesale shift in attitudes towards ecology might be
impending in the executive branch, bringing greater attention to state and local concerns with the preservation and development of natural resources.

By Fall 1970 it appeared that the lease-cancellation
approach had been abandoned by some of the defenders of
Santa Barbara's interests. Congressman Teague, for
example, introduced legislation in September[30] that
would suspend all drilling on existing leases for at
least five years, unless Congress approved renewed
drilling, and thereafter only when the Secretary of the
Interior certified that it was safe, geologically, tech-
nologically, and ecologically. The secretary would then
be authorized to extend the terms of the leases for the
period of the suspension, to waive all rentals, and to
pay interest to lessees based on their investment. A
unique note was struck in that money was authorized to
be spent for purposes of camouflaging the oil platforms
according to a plan approved by the Secretary of the
Interior "in consideration of the preferences expressed
by the residents of Santa Barbara, California."[31] This
bill also authorized expenditures for the purpose of
offshore drilling research.

     In this connection the July 1970 Interior Subcom-
mittee hearings, particularly the testimony on and sub-
stance of Cranston's bill S.1219, displayed some interesting refractions of the Nixon administration's "New
Federalism." Republican Senator Bellmon of Oklahoma,
representing a state whose capitol grounds contain
numerous oil wells, could not see why the aesthetic
qualities of another community should warrant an exception to customary practices of American development.
Moreover, he puzzled why Santa Barbarbans, if they
felt so strongly about the matter, would use oil to
drive their cars and heat their homes.[32] Democratic
Senator Cranston argued strongly that varying locales
required varying treatment by the federal government;
he emphasized "that by voting to prohibit oil development in portions of its tidelands, California has voted
to deny itself the substantial bonuses, rents, and royalties which would flow into the state treasury from
such leases. California has voted to deny itself the
state income and property taxes which this oil development would produce."[33] Senator Cranston further noted
that this action by the State of California indicated
that its citizens felt that there were some things
"simply more important than the economic advantages of
oil production under present technology."[34] And he
urged that senators with other viewpoints, who might
disagree with California's priorities, should be toler-
ant of diverse value priorities. Senator Cranston could
"see no justification for the Federal Government's denying the people of California their right to protect
their coastal environment from oil pollution."[35]

     It was with an eye to this matter of local "self-
determination," that Senator Cranston had the week
before amended his bill S.1219. As amended, the termination of the one federal lease, P-0166, would depend
upon California's termination of the California State
Lease, PRC-3150, which drew from the same wells.

     In so amending S.1219, Cranston was pointing to
an important interjurisdictional problem in the final
outcome of the Santa Barbara oil controversy. Although

neither state legislative nor state executive actions
could determine directly the nature of that final settlement, state decisions about its own resources were
likely to become factors. As numerous state legislators
argued shortly after the spill took place, the federal
government could hardly be expected to reverse the
course of development in federal waters if the state
did not impose limitations on resource utilization
within the three-mile limit. Anxiety about the prospect of federal development in very deep waters flared,
especially after Humble Oil made discoveries on leases
where the water was 1,100 feet deep. As Senator Cranston pointed out, to drill there would require building
a drilling platform the height of the Empire State
Building; a blowout or an aggravated seep at 1,100 feet
presented horrendous ecological problems. "We cannot
use divers at that depth. Even tethered divers from a
diving bell have never even attempted to operate at
these depths. The world record for such a dive is
1,000 feet. The normal maximum operating depth for
divers is 400 feet."[36]

All this might be very true. Yet none of what
Senator Cranston, or anyone else testifying against
Channel oil production said, induced the members of the
Senate subcommittee to report out any of the proposed
bills by the end of the 91st Congress. Opponents of
Channel oil development fared no better in the House of
Representatives. All proposals died when Congress
adjourned in late 1970.

The First Session of the 92nd Congress produced
largely the same empty results. As far as "the Santa
Barbara question" was concerned, it began on January 27,
1971. On that day Senator Cranston and his freshman
Democratic colleague John Tunney, who had handily ousted
Republican Senator Murphy in the November 1970 elections
introduced Senate Bill 373. S.373 lay somewhere between
the "get oil out forever" position, and the administration's compromise of the 91st session. It increased to
38 the number of leases to be included in a permanent
federal ecological reserve, and placed a moratorium

on 29 other leases until such time as technological progress made possible safe underwater completions of oil wells. Similar in intent to H.R. 2637, introduced two days later in the House by Congressman Teague, S.373 also provided for public hearings on all drilling applications, and envisaged removal of existing federal platforms in the Channel. Under both bills, Union's platforms and Sunoco's nearby Hillhouse platform were to be removed when leakage finally stopped, while those of Phillips Petroleum off Carpinteria were to be taken down when California brought an end to oil production on its adjacent tidelands.

On April Fool's Day, 1971, Senator Edmund Muskie upped the political ante by introducing S.1459, "The Santa Barbara Channel Preservation Act of 1971." That bill virtually embodied the desires of those most hostile to oil in the Channel. It sought to cancel all Channel leases in federal waters except those with existing platforms. The operation of these platforms would be transferred from the private oil companies to the Department of the Interior with instructions: (a) to remove them when it was safe to do so; and (b) to settle any damage claims which should arise from lease cancellations with revenues derived from other oil-bearing federal lands.

A few weeks later four Republican Senators--Gordon Allott, Wallace F. Bennett, Jacob K. Javits, and Ted Stevens--countered with the administration's more modest proposal raising to 35 the number of leases to be placed in a national energy reserve, but providing for no moratorium on the remainder.

Introduction of bills, however, was about as far as matters went in Congress for the rest of 1971. As late as January 1972, when the 92nd Congress reconvened for a second session, no bill had even gotten a committee hearing.

The 1971 Cranston-Tunney bill reflected in one of its provisions a fact of political life in a federal system--one with which opponents of Channel drilling have needed to reckon from the start. By conditioning the termination of oil production from Phillips' federal platform on prior termination of California Standard's production in adjacent state waters, S.373 points to the circumstance that in order to succeed in one jurisdiction opponents of drilling may need first to persuade the government of another--and *vice versa*.

THE ROLE OF THE STATE

This need may sometimes pose a well-nigh insuperable obstacle to desired change. Yet, in the Santa Barbara case, advocates of change had at least one advantage. The state had already tried to do something. As Senator Cranston had noted in his July 1970 subcommittee testimony, even before the spill, California had not been inactive in attempting to protect its coastal tidelands. In 1955 the State Legislature had created sanctuaries safeguarding almost all of Southern California. The largest sanctuary stretched

> from Newport Beach down the historic
> coast of Orange and San Diego Counties
> to the Mexican Border. The offshore
> Islands of San Clemente and Santa
> Catalina were set aside as were the
> heavily used Metropolitan Beaches
> of Los Angeles and the magnificent
> sixteen-mile stretch of Santa Bar-
> bara Beach....The rugged and un-
> spoiled coast of San Luis Obispo
> County completed the 1955 actions.
> In 1963 two additional Northern
> California coastal areas [were set
> aside]...Monterey Bay....as well as
> the desolate grandeur of the Big
> Sur...and the wild, redwood lands
> of Humboldt and Mendocino Counties.[37]

After the spill, the State of California was by no means
wholly supine. The Monterey Bay sanctuary was extended
north to the border of the adjoining county, and at
present these sanctuaries account for almost one-quarter
of the whole California coast. Moreover, "they prohibit
oil drilling on an estimated half of the tide lands
suspected or known to contain oil deposits."[38]

During the 1969 and 1970 California legislative
sessions, numerous bills on water pollution were proposed specifically relating to the Santa Barbara conflict. By early 1970, however, party politics were becoming intermingled with the issue. Most notably, Jesse
Unruh, then State Assembly Minority Floor Leader and
the 1970 Democratic gubernatorial candidate, argued for
a bill to bar all state drilling offshore. Santa Barbara's Republican Assemblyman, W. Don McGillivray, almost simultaneously introduced a bill which sought to
establish a moratorium, rather than an outright ban.
The Santa Barbara Democratic Central Committee charged
that the McGillivray bill was not "militant" and produced a rejoinder from the GOP Santa Barbara Central
Committee.[39] As of late June 1970, the most that was
achieved by the Assembly on this issue was the passage
of a watered-down McGillivray bill. Having initially
prohibited new leases in state waters, the policy was
diluted to allow drilling where, in the eyes of the
State Lands Commission a lease would not result in
"seepage or spillage of oil or in the destruction of
scenic or aesthetic values."[40] It would be rather difficult for the State Lands Commission to be sure that a
lease would not have any of these unhappy results, thus
limiting the bill's significance. Moreover it did not
go as far as Unruh would have chosen, nor as far as
liberal Republican Assemblyman, William T. Bagley,
wished. He introduced another bill seeking to bar all
drilling from Monterey, 80 miles south of San Francisco,
to the Oregon border.

Two actions undertaken by the state since the
spill and up to July 1970, were more significant.
First, the State Lands Commission at its first meeting

after the blowout on Platform A imposed a moratorium on all existing state leases. As Houston I. Flournoy, State Controller and Chairman of the Commission, stated in his testimony at the second of the United States Senate Minerals, Materials and Fuels Subcommittee's Santa Barbara Hearings, on March 13, 1970: "We've not allowed one new well to be started on any existing leases anywhere within the California tideland under the State Lands Commission jurisdiction--whether by onshore slant drilling or offshore operations...."[42] Flournoy urged that the federal government adopt a similar moratorium on new drilling with the exception of the operations near Platform A. He drew attention to the First Report of the President's Panel on Oil Spills which stated, "the United States did not have at this time sufficient technical operational capability to cope satisfactorily with the large scale oil spill in the marine environment."[43] Flournoy concluded his testimony with a pointed comment: "We believe that many of the recommendations of the President's Panel on Oil Spills are meritorious and of at least equal import to the recommendation of the DuBridge Report for remedial operations at Platform A. We only wish that they were being implemented with equal vigor."[44] This point of view, arguing for state and federal parity, was directly in line with Senator Cranston's amended version of S.1219 four months later.

One other significant action was undertaken during the 1970 state legislative session. Under the leadership of Santa Barbara's State Senator, Robert J. Lagomarsino, who was Chairman of the State Senate Committee on Water, Wildlife and Natural Resources, two resolutions to President Nixon and the United State Congress were approved. The first called on the federal government to match the existing state moratorium, and the second asked it to enact a federal equivalent to the state sanctuary. Both passed on February 11, 1970: in the Senate by a vote of 28-0, and in the Assembly by 64-0.[45]

These two actions--one by the State Lands Commission and one by the State Legislature--were the most significant California actions concerning the future of the Channel undertaken during the first two years after the spill. They suggest a further problematic aspect of Santa Barbara's efforts to achieve federal redress of grievances through state action: the political role played during the controversy by the state's chief executive. Although certainly the California governor could not simply order the oil companies out of federal waters, to many Santa Barbarans there was little doubt that he could profoundly influence the course of events in at least two ways. First, the governor could take a strong stand in proposing or supporting legislation to protect California's coastal ecology. As Perry Davis, Public Affairs Manager for the Los Angeles District Office of the Army Corps of Engineers, remarked in January 1970, the "Corps would have to view much more seriously a State protest than it has so far considered the City and County objections."[46] One may reasonably assume that the Army Corps was not the only federal agency or branch disposed to such a view.

This suggested the second path of possible gubernatorial influence--pressure directly upon Washington through the inherent stature of his office and his influence in the national Republican Party. Governor Reagan's degree of activism was distinctly less than many Santa Barbara observers desired during the months following the spill. And, although Lois Sidenberg of GOO reported in January 1970 that the governor seemed more interested than he had appeared earlier, and although Houston Flournoy stated at the May 1970 U.S. Senate Interior Subcommittee hearings that he spoke not only for the State Lands Commission but also for the governor, it would be premature to conclude that all local doubts on this score have been dissipated.

On the contrary, probably doubts have increased during the past year. Political behavior in Sacramento during 1971 suggested that the state government may not be the place where those wishing a lever to produce

federal legislative action should place great hopes.
What emerged from the 1971 California legislative session was largely unsatisfactory both to those specifically concerned with Channel drilling and to those
anxious for far-reaching general environmental legislation. Most of the important ecologically oriented
bills introduced into the State Legislature died in
committee--including a bill to create a "super-agency"
with veto power over the activities of major state agencies charged with responsibilities affecting the environment. Another significant bill that died for want
of a majority vote to report it out of a Senate committee was A.B. 1471: Passed by the lower house, A.B. 1471
would have stemmed developmental activities along the
1100 miles of California coastline unless they secured
approval by a coastline regional commission.

Even if a bill passed both houses it still had to
get by the critical eye of the state's chief executive.
And, in 1971 that eye was more critical of positive
legislative output than ever before. By Christmas 1971
Governor Reagan had set the all-time California record
for gubernatorial vetoes, surpassing in this respect
even California's reforming governor of the early
twentieth-century, Hiram Johnson. Only one really important environmental bill "slid through." S.B. 789,
introduced by State Senator Lagomarsino, created a new
power for the state attorney general to pursue the public
interest in protecting the environment by instituting
court actions. But, it should be noted, from the perspective of at least some environmentalists, that victory was of dubious value. By placing his signature on
S.B. 678, Governor Reagan effectively dissipated the
chances for passage of other bills in the legislature
that sought to open up a broader front for environmental litigation--permitting individual private citizens
to initiate environmental suits in California courts
without demonstrating personal injury.[47]

On balance, by early 1972 Santa Barbara foes of
Channel oil development had no good reason to be
entirely satisfied with the performance of either state

or federal legislatures. Indeed, their primary success came from the federal executive. Moreover, that success gave promise of being only short-lived. Secretary of the Interior Rogers C.B. Morton, even as he had added 15 new tracts to the 20 parcels on which his predecessor had declared a development moratorium, had indicated that unless Congress passed some statutory restrictions the executive moratorium would cease on January 1, 1973.

CHAPTER **VII**

Legal Remedies

If difficult cases make bad laws, the Santa
Barbara oil slick may well produce numerous legal as
well as ecological ill effects. By the end of October
1969, Judge Albert Lee Stephens, Jr. of the Federal District Court for the Central District of California, had
before him some 20 cases arising out of the spill.[1]
Because of the abundant litigation and the unsettled
nature of the relevant law, we shall not attempt here
to judge each present and potential case upon its
merits. Rather we content ourselves with delineating
the underlying policy perspectives that will come into
play during their argument; with mapping out the apparent barriers facing plaintiffs; and with sketching possible strategies for circumnavigating these barriers.
We turn first to the policy views likely to underlie
the oil industry's and the federal government's defense.

THE DEFENSE: THE DAMNUM ABSQUE INJURIA
PERSPECTIVE AND SOVEREIGN IMMUNITY*

From one perspective, John Marshall was at best
naive when he assumed in the primeval days of *Marbury*:[2]
"The very essence of civil liberty certainly consists
in the right of every individual to claim the protection
of the laws, whenever he receives an injury."[3] Despite

Damnum absque injuria=an injury without a legally
recognizable wrong.

what the great Chief Justice thought, it was not always "one of those first duties of government...to afford that protection."[4] On the contrary, no obloquy whatsoever was cast upon American jurisprudence because it withheld remedy in numerous cases.

Remedy Withheld in the Interest of Progress

This is of course the *Rose v. Socony-Vacuum*[5] view of sensible jurisprudence, as the law relates to American economic growth. The doctrines of sovereign immunity and *damnum absque injuria* both frequently operate to fulfill a primary function of the law. That function is to ensure a sufficient vacuum of freedom from responsibility for the inadvertent side effects of industrial activity--a freedom from responsibility sufficient to ease the accumulation and use of capital for purposes of economic growth.

This view of industrial things may, to be sure, overlook the smog suffered by nearly every megalopolitan resident. Nonetheless, it has the advantage of removing what looks like a legal logjam by pronouncing it to be nothing of the sort but rather a soundly constructed dam against interference with American progress. In American jurisprudence this view wears a century's aura of respectability,[6] and we can anticipate its future use by attorneys for both oil companies and the federal government when confronted with ecologically based claims. Let us recapitulate the lineage of this view.

If former Supreme Court justices had been able to hear the Santa Barbara controversy, it is doubtful that any majority since the era of the Taney Court[7] would find a "crime" in Union's mishap. In the days of Taney when the sea's oil came only from whales, and when judicial dispositions bent strongly towards permitting states to police "enterprise" as they might,[8] two prerequisites for allowing recovery against Union might have been met. First, sufficient utility might still have been found in the older eighteenth century legal view that a wrong required *ipso facto* reparation.[9] Second, state liability policies might not have been found to be subordinate to national considerations.

Social Darwinism

The reason why later courts, at least through 1936, would have found no remediable damage is plain. The ideologies of Social Darwinism and laissez-faire legitimized the need for capital accumulation during the takeoff stage[10] of American capitalism and justified not holding the entrepreneur to a close accounting for the noxious side effects of his enterprise. So, too, during the drive to economic maturity.[11]

If the fittest are to survive, the risks of being fit ought not be too awesome or too expensive. Otherwise the fitness competition might not have sufficient participants to ensure national economic growth. Therefore, a shift occurred during the second half of the nineteenth century to fault liability in the law of torts, thus moving the burden of proof of negligence to the plaintiff's shoulders and away from the otherwise beleaguered entrepreneurial defendant. The concept of no liability without fault became the prevailing attitude. And since accidents do sometimes happen, despite the exercise of reasonable prudence, the doctrine of *damnum absque injuria* prevailed as well.

A British Reservation

In England--the common law's original source and then the world's industrial leader--Victorian courts were subscribing to a jurisprudential shift similarly geared to lessening the entrepreneur's anxiety about liability. But simultaneous to the election of Ulysses S. Grant as national chief executive and *maître de* of the "Great American Barbeque,"[12] Britain's highest judicial forum, the House of Lords, attached an

important limit to the emerging doctrine. That limit
was set by the Law Lords' opinion of 1868 in *Rylands v.
Fletcher*,[13] which quickly became a major touchstone for
plaintiffs on both sides of the Atlantic angry about
the damaging externalities, the injurious side-effects,
of nearby industrial activities.

In *Rylands*, the defendant had dug and filled a
water reservoir. The bottom of the reservoir was next
to some long abandoned mine shafts which, unknown to
him, were connected with a newer mine being worked by
the plaintiff. Once filled, the reservoir inundated the
new mine with water. His business brought to an
untimely end, the plaintiff filed a tort suit against
the defendant. The defendant argued *damnum absque
injuria*: He had not been negligent; therefore he had
committed no tort--no cognizable wrong. It was just
an unfortunate accident without a legal remedy. Not
so, determined the British judiciary. When engaged in
an ultrahazardous activity, the entrepreneur must observe the maxim *sic utere tuo ut non alienum laedas*:
"so use your property as not to harm others."[14] Con-
structing water reservoirs was an extrahazardous use of
property, consequently even in the absence of negligence the defendant would have to pay up.

Eighteen years later in what was to become *Rylands'*
"American challenger,"[15] *Pennsylvania Coal Company v.
Sanderson*,[16] a bare four-to-three majority of the
Quaker State Supreme Court declared:

> It must be conceded, we think, that
> every man is entitled to the ordin-
> ary and natural use and enjoyment
> of his property....If, in the ex-
> cavation of his land, he should
> uncover a spring of water, salt
> or fresh....he will certainly not
> be obliged...to conduct it out of
> its natural course, lest the
> stream in its natural flow may
> reach his neighbor's land....He

> may, to a reasonable extent, *jurae naturae*, divert water from a stream....[17]

In the instant case, the Pennsylvania Coal Company could, *jurae naturae*, render useless Mrs. Sanderson's drinking water, pipes, dam, and fish pond by discharging acidulated water--the "natural by-product" of its mining operations--into a brook which flowed downstream onto her property. It could do so despite the fact that that the brook's purity had been a major reason for her purchasing and improving downstream property. In the court's majority opinion, there was nothing unreasonable in the coal company's activity, despite the fact that, on two earlier hearings of the same case,[18] the same bench--with a slightly different staff--had relied on the *sic tuo utere* maxim and held the coal company liable. In reversing itself, the court might have felt "embarrassed...because of the well-known ability and learning of the distinguished judge who delivered the previous opinion, and of the fact that two, at least, of our number have given that opinion their formal approval."[19] Moreover, the 1886 Court might have found some difficulty in clearly determining how far *Fletcher v. Rylands* was honored in other jurisdictions.[20] It might have seemed to be straining somewhat in distin-- guishing *Sanderson* from a lengthy chain of earlier Pennsylvania cases that appeared to hold the upper riparian owner responsible for pollution to downstream property.[21] Nevertheless the majority had no difficulty in deciding that coal mining was crucial to the economic development of Pennsylvania,[22] and that there was "scarcely a stream in the mining regions of Lackawanna County...not, to a greater or less extent"[23] polluted by mining operations. The balance of interests at stake was clear:

> The plaintiff's grievance is for a mere personal convenience...that... must yield to the necessities of a great public industry, which, although in the hands of a private

> corporation, subserves a great
> public interest. To encourage
> the development of the great
> natural resources of a country,
> trifling inconveniences to par-
> ticular persons must...give way.[24]

In 1886, the "Wabash Year,"[25] the Pennsylvania
judges played rather more nearly the legal tune of their
federal superiors, more so than in the days when their
famous forebear, John Gibson, attempted to make short
shrift of Marshall's theories of judicial review. He
later failed "promotion" to the Supreme Court for that
reason.[26] In any event industry was viewed as producing annoyances, without legal wrongs. Thus legal responsibility for pollution became more and more difficult to assign. Much to the advantage of corporate
enterprise in the Age of the Barbeque, the retreat of
the courts paralleled the disappearance of legislative
responsibility from the congressional floor to the sha-
dows of committee chambers.[27]

What has this to do with contemporary pollution
and, specifically, offshore oil pollution? By 1937--as
conventional wisdom would have it--the legal balance of
economic interests had changed. After 50 years of con-
fusing constitutional law with Herbert Spencer's *Social
Statics*,[28] the Supreme Court abruptly ended its affair
with Big Business and once more allowed rule-makers to
legislate entrepreneurial limitations.

Liability Still Unclear

This may be true in general. But in four respects
liability for damage resulting from offshore drilling
in federal jurisdiction remains far from clear.

First is the curious American adherence to the
colonial era mode of redress against the Crown. Thus
the United States, after 1776, maintained the doctrine
of the sovereign's immunity from suit--merely

transferring the immunity from Crown to government.
However, disappearing in the shuffle with Cornwallis'
retreating armies was the doctrine's offsetting counter-
part--the subject's right to petition the monarch, a
mode of redress granted, *de facto*, as a matter of
course. And, oddly, no American equivalent was then
created in its place.[29]

The second reason lies in judicial insistence upon
a narrow interpretation of the Federal Tort Claims Act
of 1946.[30]

Third is the oil industry's successful application
of pressure upon Congress in 1966 to limit its liability for damages from spills to "gross negligence,"
which is even less responsibility than that ordained by
congressional policies in the 1924 Oil Pollution Act
passed in the "Age of Normalcy."

Fourth, and ironically, in view of liberal-
conservative attitudes prevailing during the tidelands
jurisdictional controversy from 1945 to 1953,[31] the
Supreme Court has helped, not hurt, industry. The 1947
opinion in *United States v. California*, "the First,"[32]
may have seemed inconvenient to companies anticipating
that federal control would restrict the terms of production and opportunities for profit more than would
establishment of state jurisdiction.[33] But the Submerged Lands Act of 1953 undid that opinion. Moreover,
the 1965 opinion in *United States v. California*, "the
Second,"[34] which disavowed California's claim that the
Channel was a state-owned bay, is eminently convenient
for the industry now.

If this view of legal matters is widely accepted,
then what chance is there for those who seek restitution for damages in the Santa Barbara oil spill beyond
that which Union is willing to make without a legal
battle?[35] First, on the court's understanding of the
Federal Tort Claims Act, ample sovereign immunity remains to bar recovery against the United States.
Second, the state courts are foreclosed from hearing

litigation against Union by section 1333 of the Outer
Continental Shelf Lands Act, which gives jurisdiction
to the Federal District Court for the Central District
of California. Third, the 1966 amendment of the Oil
Pollution Act reflects a congressional policy of
"laissez-drill." Consequently, section 1333 of the
Outer Continental Shelf Land's Act (OCSL) tells
the Federal District Court generally to apply local
state laws "to the extent that they are...not inconsistent with this Act or with other Federal laws and regulations of the Secretary now in effect...."[36] Clearly,
then, California's antipollution laws[37] which impose
penalties without regard to degree of negligence are in
conflict with federal laws. Further, judicial application here of California's absolute liability doctrine[38]
to the oil industry would be equally inconsistent.
Unless it can somehow be proved that Union was grossly
negligent in its operation of Platform A--Union Oil is
not liable. Federal policy overrides: The doctrine of
damnum absque injuria prevails over the maxim *sic utere
tuo ut non alienum laedas*.

OBTAINING SATISFACTION: POSSIBLE ROADS TO REDRESS

The foregoing logical path to *damnum absque injuria*
will seem crooked to the State of California, its South
Coast county and municipal subdivisions, and other present and future plaintiffs.[39] It is not hard to imag-
ine at least four avenues of alternative policy that
these plaintiffs may choose to argue. First, they may
urge that the country needs now, as never before, to
focus upon the threat of pollution. Consequently it
is desirable in the name of cooperative federalism to
encourage the activities of those states, and their
local subdivisions, that genuinely attempt to function
as experimental workshops in combatting pollution.
Second, the plaintiffs may argue that the time has
long passed when the national interest required codd-
ling the infant oil industry. Like aviation, oil has
passed through the stage of necessary natal permissive-
ness, and the time for noting its deleterious ecological side effects has arrived.[40]

Third, a more general policy line of reasoning may
be advanced about the relationship between law and the
economy. If the nineteenth-century economy required
the law's departure from strict responsibility in order
to assist the economic takeoff, the mature late-twentieth
century economy of "countervailing powers"[41] requires
either shifting to absolute liability[42] or jettisoning
the whole concept of "wrong-doing" in favor of allocating the costs of remedies according to the ease of bearing the burden.[43] Such a shift, it might be urged, is
particularly desirable when one party is too small to
"countervail" the other in the political arena. Here
the courts have a special political duty to keep open
the paths of redress. And finally, the plaintiffs might
argue that the foregoing history of *damnum absque injuria*
is drastically oversimplified.[44]

None of these policy propositions contains such
inner contradictions that its mere statement suffices
to refute it.[45] However, speedy judicial determination
of the conflict appears highly unlikely.[46]

Three Plaintiffs' Problems

Present and future plaintiffs have three major
problems to resolve in pursuing litigation: (1) the
available forums; (2) the optimum types of suit possible; and (3) the most favorable substantive law likely
to be applicable.

Finding a forum. Two advantages would be open to
a plaintiff securing a state forum, either in a criminal or civil suit: the greater chance of finding a
favorably disposed judge; and the greater ease of applying California law, which is less favorable to the oil
companies' cause than whatever federal law may exist.[47]

However, such a forum may not be available.[48]
There is sufficient ambiguity in the law to present the
threat that the defendant may remove the case to a federal court. This difficulty has already been manifested

by the temporarily successful effort of the oil industry
to secure a federal injunction against Santa Barbara
District Attorney David D. Minier, who threatened to
prosecute in state court for public nuisance. Similarly, the state attorney general's office has also indicated uncertainty on this score. After an initial
inclination to file its $560,006,000 damage suit in
federal court, a decision was made in mid-February 1969
to file in the state court in hopes of a more favorable
forum. A few weeks later, after an application by the
oil companies for removal to the federal district
court, the state decided to file *de novo* in the federal
judicial system.[49]

The central problem for any plaintiff seeking to
use state courts is contained in that troublesome section 1333 of the OCSL Act of 1953. Does the phrase,
"giving to the Federal District Court original jurisdiction in suits arising out of offshore leasing" mean
to give *exclusive* jurisdiction to that court? Certainly
it cannot entirely foreclose all other forums. Surely,
for instance, the State of California has constitutional
standing to invoke the original jurisdiction of the U.S.
Supreme Court in a suit against a "suable" officer of
the United States.[50] Probably more important in terms
of possible frequency, does this clause bar all concur-
rent exercise of jurisdiction by state courts even after
the damage has reached property on California shores?

Types of possible suits. A potential difficulty
from the plaintiff's standpoint of giving up on the
state forum is the apparent hindrance to the pursuit of
the second of the two major forms of relief available--
suits for damages and abatement of nuisance. Undoubt-
edly, a state forum could find sufficient "continuity"
in over 500 days of "spill," with no apparent end in
sight, to constitute a public nuisance.[51] Failure to
comply with an order of abatement would subject Union
to criminal penalties. Such is the argument of Santa
Barbara District Attorney, David Minier, and the reason
for his argument against exclusive federal jurisdiction
in seeking a state court holding against drilling.[52]

It would probably be difficult to persuade a federal
prosecutor to undertake a similar action in federal
court. But there should be a way around this--as long
as the requirement of continuing damage is met. Could
not an individual citizen of California follow William
Prosser's recent "advice"[53] and seek such an injunction
in the federal district court?[54] Whether he would be
successful is, of course, another matter. On balance,
suits for damages appear to be more likely to bear
fruit.

   Applicable substantive law. If a plaintiff had
decided in February 1969 that litigation in federal
jurisdiction was the better course, and that a damage
suit was more "hopeful" than injunctive relief, he
would still have had to weigh the advantages and disadvantages of appearing before the federal district court
sitting in admiralty. Leaving aside the potentially
conflicting scope of the amended 1924 Oil Pollution Act
as indicative of congressional policy in 1969,[55] there
was but one clear procedural disadvantage as against
three substantive advantages.

   The single disadvantage was that the plaintiff
would have to start all over again if he were denied
admiralty jurisdiction. This indeed constituted a problem, compounded by the unsettled nature of the law
where damage occurs on land as the result of an occurrence at sea, or *vice versa.*[56] Consequently, a beachfront property owner might well have decided that it
was better to avoid admiralty jurisdiction. However, a
boatowner might well have reasoned differently: If a
swimmer struck by a surfboard can recover for an injury
in admiralty, might not a boatowner also anticipate
reparation for hull damages?[57] As a general rule admiralty jurisdiction offers two benefits: Contributory
negligence is not a barrier to recovery,[58] and there
appears to be no admiralty doctrine of "assumption of
risk."[59] At most, a doctrine of comparative negligence
might result in apportioning the damages.[60] In addition, a third more specific advantage might emerge: A
ruling that a rig was a vessel and hence subject to the

doctrine of "unseaworthiness" would open up the road to broad recoveries.

Had a plaintiff so chosen in February 1969, he would would not have reaped these three substantive advantages four months later, but rather the overriding procedural disadvantage of starting all over again. On June 8, 1969, in a Louisiana offshore drilling case, the Supreme Court decided that platforms do not fall within admiralty jurisdiction.[61]

Possible Results of the Federal Civil Route

Thus, the route of federal civil jurisdiction, which the State of California and the County and City of Santa Barbara had selected, turned out to be the wiser one. There are essentially three possible substantive results of pursuing this route.

Federal adoption of California civil law in its entirety. The most advantageous outcome for the plaintiffs would be to persuade the federal courts that California civil law is applicable. In recent decades, California courts have been favorable to doctrines of absolute liability, at least in respect to ultrahazardous activities.[62] Even as long as 40 years ago *Green v. General Petroleum Corporation*[63] held the doctrine applicable to oil in no uncertain terms. Later adjudications do not seem to have affected *Green*'s authority. The advantage of such a holding is, of course, that plaintiffs would not have to prove negligence in any form in order to collect. For reasons noted below[64] the wholesale adoption of California law would bar defenses based on acts of God or *vis major*. It would also--at least in regard to plaintiffs owning property proximate to the bench--bar a defense based on "foreseeability":[65] Defendants would not be able to escape liability by arguing that an ordinarily prudent man could not be expected to foresee the damage.

What are the potential barriers to the adoption of such a position? Clearly, they would have to be built upon the saving clause of section 1333 of the OCSL Act. However, the disparity between the 1924 Pollution Act as amended in 1966 and California's absolute liability doctrine is by no means crystal clear. This is so far three reasons. First, the 1966 oil spillage clause amends a purely criminal statute. It is not obvious that the penumbra of such a statute should extend to civil remedies. If the oil companies spent so much money exploring the geological characteristics of the Channel to be sure that offshore drilling was prudent then it would seem that they must have retained attorneys familiar with California law. Second, the 1924 Oil Pollution Act stated: "This Act shall be in addition to the existing laws for preservation and protection of navigable waters and will not be construed as repealing and or modifying or in any manner affecting the provisions of those laws."[66] Third, the act applied only to tankers and other vessels. It had not yet been judicially determined whether any types of offshore drilling rigs were vessels.[67] Yet, although offshore rigs were not numerous in 1924, by the time of the 1966 amendment they had long been on the scene. Consequently, it is difficult to maintain the view that congressional silence in 1966 on the matter of rigs indicated a desire to extend to them that act's coverage.

A middle road of negligence? This does not entail that a holding of absolute liability will necessarily result. By 1969 there were numerous other congressional laws on pollution.[68] And, while these laws displayed congressional concern about water pollution dating back at least to 1888, they do not expressly specify absolute civil liability. Consequently, so the argument might run, recourse must be through federal judicial decisions concerning pollution. Without discussing all circuit decisions, it seems plausible to anticipate heavy judicial reliance upon a 1923 circuit decision, *Sussex Land and Livestock Company v. Midwest Refining Co.*[69] In that case, oil escaped from the defendant's field into a river and was carried downstream onto the plaintiff's property. Viewing the result as a trespass, the appellate court rejected the *damnum absque injuria* doctrine.

In the event of conflict between the relative applicabilities of *Sussex* and cases of percolating underground waters, such as *Rose v. Socony-Vacuum*, it seems reasonable to suppose that a showing of negligence would establish Union's liability.[70] While current and wind action did not inevitably cause the blowout on Platform A to produce a "trespass" on California land, it would be difficult to sustain the view that a prudent man--or corporation--would not have reasonably anticipated such an eventuality.

The essential questions then would seem to be: To what extent should Union have foreseen the possibility of damages, and at what points, if any, was Union negligent? At least, the possibilities for proving negligence--particularly if an inference from *res ipsa loquitur* ("the thing speaks for itself") is permitted-- should provide abundant opportunities for litigation.[71] On the matter of foreseeability, it should prove interesting to see to what extent arguments for additional caution, in view of the nature of the Santa Barbara economy, would extend Union's liability beyond earlier cases. Such cases have seldom covered community losses described previously.[72]

Generalizing on the 1966 amendment. The third possible federal judicial outcome is, of course, a decision that the 1966 amendment to the 1924 Oil Pollution Act indicated congressional intent both to preempt the field and to allow drilling on any terms deemed wise by the Secretary of the Interior. We will not attempt to judge whether such a holding is likely, but would suggest, that it could hardly be expected to aid the cause of cooperative federalism.

CHAPTER VIII

Where Do We Go From Here: The Ecology, The Economy and the American Political Future

THE IRREVERSIBILITY OF OIL DEVELOPMENT DECISIONS

The still-continuing Santa Barbara oil spill has produced a political conflict between the great and the small. On one side stands the world's most powerful industry, and on the other, the "restless natives" of a small city situated between the bronze mountains of California's Coast Range and the blue Pacific. There is no reason to expect a settlement that would genuinely appeal to both sides: Transcending the question of profit distribution, the real issues are rooted in broad conceptual differences over how the political system should resolve interest-group clashes.

It would be an exaggeration to depict the Santa Barbara oil conflict as having "radicalized" a status-quo-oriented community.[1] But it would be equally inaccurate to suggest that the conflict is merely another routine dispute over "who gets what, when, and how." What makes this dispute different is that the issue of Channel oil development, once decided, will not permit further public debate and political struggle. This is true, whether the outcome is decided primarily within the sealed-off environs of an administrative agency, or principally in the even-more-invisible compartments of congressional committee executive sessions.[2]

What is important above and beyond the political norm is that the conflict has an aspect of finality and irreversibility, at least from the viewpoint of the

Santa Barbara South Coast community. A large number of
gigantic oil platforms in the Channel threatens the
Santa Barbara economy and environment. The spill is
not analagous to, for instance, a wage dispute between
the United Auto Workers and General Motors, which is
part of a continuing process of marginal adjustment.
Once the die of development is cast, the future of Santa
Barbara is determined.

Many political scientists view the Santa Barbara
oil conflict as different from those the American political system has so successfully solved in the past.
Until such time as truly dependable underwater drilling
and well completion may become possible, and a greater
technological capacity to remedy damages from spills
can be achieved, it is unlikely that all the government's
men could put Santa Barbara's ecosystem back together
again. In this sense, the post-New Deal pattern of
governmental "partial decision-making,"[3] which Theodore
Lowi has brilliantly criticized, might well turn out to
be "partial" in a rather different sense from what Lowi
means. However the decision is reached, it is bound to
be *partial* or biased, in the sense of favoring one of
the two contending factions. But if it favors the oil
industry, it becomes final and virtually irreversible.
Any prognostication about the ultimate settlement of
the controversy should be framed in light of these observations. We offer two alternate forecasts of the
problem's likely solution, taken from the evidence accumulated since January 28, 1969.

THE "PRO-OIL" OUTCOME

A sensible pessimist would concede the possibility
of a particular miracle. That is, he could not easily
deny that the President might be visited by the ghost
of John Muir or Gifford Pinchot. Accordingly, the President could decide that the national interest no longer
truly required exploitation of the Santa Barbara Channel's oil resources, but necessitated termination of
the leases, and cessation of development, with just

compensation for the oil companies' investments. It would not matter whether such compensation came by a mutually agreeable exchange for oil-bearing lands else-where--such as in the federal Elk Hills Reserve[4]--or by direct monetary compensation. A presidential decision to eliminate all but the present Union Oil and Sunoco drilling to deplete the pressure in the Dos Cuadras field, coupled with active Interior Department partici-pation in maximum-feasible application of "sealing-off" technology, would almost certainly overcome any resis-tance in the Congress. Similarly, a particular con-gressman from Arkansas, the Chairman of the House Ways and Means Committee, might possibly decide that the tax depletion rate had to be reduced to the point of making further Channel drilling a marginal investment.

But, from the pessimist's viewpoint, such outcomes are unlikely. Moreover, such outcomes would not affect his general understanding of how distributory disputes are settled in the American polity. Thus he might argue along the lines of a critic of pluralism such as Theo-dore Lowi. He might hold that the liberal-conservative dialog no longer has significance, but that a pro-industry outcome is virtually preordained.[5]

In substantiating this probable predisposition to-ward a pro-oil settlement, a pessimist would only need to look at what has so far happened by way of execu-tive, legislative and judicial actions involving the competing interests and values.

The Executive Branches

He might well see policies of the executive branches of both state and federal governments as grist for his mill. It would, indeed, be difficult to argue that the chief executive of California has exerted any great ef-fort on behalf of the cessation of Channel oil drilling. One could point to his relative lack of initiative in proposing state legislation that would put pressure on the federal government. Also noteworthy is the fact

that he has not--publicly at least--stated directly
to the federal chief executive, as one Republican
leader to another, "we need an end to drilling."

Despite the October 1969 recommendation of the
White House panel on oil spills that public hearings be
held prior to issuing new drilling permits,[6] the chief
executive permitted the Army Corps of Engineers to continue in the old secretive ways. Pressure in federal
court in the form of county, city, and ACLU suits seeking public hearings produced no speedy, major effect
upon the White House. Similarly, the recommendations
of both a presidential task force and the cabinet majority to overhaul the oil import system in favor of tariffs did not receive quick support from the White House.
The pessimist might argue as did Houston Flournoy before the Moss Interior Subcommittee, that the executive
agencies of the federal government should follow the
advice of the President's first oil-spill panel--stress-
ing the inadequacy of present remedial technology--with
a vigor equal to that shown in following the recommendation of the DuBridge Commission, that Union Oil be
permitted to expand oil extraction in order to reduce
oil pressure in the Dos Cuadras field.[7] Yet this did
not occur. Speedy publication could have been expected
of the testimony and reasoning behind the DuBridge Commission's recommendations--if, that is, the weight of
its testimony shows that the advantages of pursuing
these recommendations overbalance the dangers of subsi-
dence resulting from drilling, as outlined by one critic
of the report.[8]

On this view of the matter, it is difficult to regard the positive executive actions emanating from the
White House and the Department of Interior as much more
than half-hearted. A priori, one might well have
thought that the President would perceive the oil controversy as Senator Cranston described it: an excellent
opportunity for applying the decentralizing principles
of Mr. Nixon's oft-enunciated "New Federalism."[9] There
have been six executive actions bearing specifically
upon the Channel oil situation to date: (1) the

tightening of drilling regulations ordered by former
Interior Secretary Hickel in spring 1968; (2) the creation of the Ecological Preserve and the "new buffer
zone," (3) the recommendation, embodied in S.4017 and
introduced in the Senate by then Senator Murphy on June
24, 1970, that 20 of the 71 Channel leases be terminated; (4) increasing one year later to 35 the number of
leases to be so terminated; (5) holding public hearings
in early 1971 on both Sunoco's and Union's applications
to erect new platforms on already producing leases, and
(6) denying those applications in September, 1971.

As was pointed out in a June 1970 editorial in the
Los Angeles *Times*, the original administration bill did
not even go as far as had been advised by the Council
on Environmental Quality.[10] Just months before, CEQ
had recommended that at least 11 tracts near San Miguel,
one of the Channel islands, be added to the 20 leases
to be withdrawn.[11]

Testifying before the Moss Subcommittee on July 23,
1970, Secretary Hickel pointed out that the administration's cumulative actions since the spill had in effect
withdrawn from any proximate threat of exploitation
198,000 acres seaward of the Santa Barbara State Oil
Sanctuary. For Secretary Hickel, these actions signi-
fied "that we have reached that state of historical re-
finement where--in some areas--we willingly trade financial for nonmonetary returns, knowing that the long-term
benefits to society exceed by far the short-term economic gains...."[12] Santa Barbara might well

> be the hallmark....where the progress
> of America is measured not solely by
> the rise in personal income, or gross
> national product, but also by the
> miles of clean beach, the acres of
> wildlife habitat, and recreation areas
> we can enjoy and preserve for future
> generations.[13]

What then of the decision the following April to add another 15 leases to a permanent statutory sanctuary? The answer from the pessimist's point of view would almost surely be: But this still leaves half of the Channel acreage open for exploitation.

Moreover, scrutinizing the Army Corps of Engineers, an observer would have difficulty perceiving any great shift in their techniques of balancing social benefits and disbenefits, particularly in their insistence that only the Interior Department can have concern for aesthetic factors in allocating drilling permits. Thus far, the matter of Army permits could not be considered encouraging. As late as June 1970, the Army corps informed Lois Sidenberg, President of GOO, that Standard Oil of California soon would begin exploratory drilling north of Anacapa, another Channel island. The corps had permitted this despite the fact that parts of the two tracts involved, #353 and #361, lay in the Coast Guard's designated westbound sealane.[14]

The foregoing might be evidence that the executive branch remains partially deaf--particularly at lower levels--to the interests of constituencies other than those it supposedly regulates, in statutory theory if not in bargaining practices. With respect to the substitution of coherent planning for *ad hoc* "program-making," the prospects do not yet seem highly promising. Daniel Moynihan, Special Counsel to the President, argued for this shift in the Summer 1970 issue of *The Public Interest*.[15] But at almost the same time the White House National Goals Research Staff suggested that such a shift in executive leadership was nowhere near imminent, since the Presidency was not, in its view, the place for settling national priorities. The report left that more or less "to the people."[16] This was a long way from the duties of "good conservative" leadership once suggested by an analyst as characterizing the other principal conservative party in the Anglo-Saxon world, the British Tories: "skirmishing far ahead of the rank and file."[17] The White House National Goals Report did not come close to suggesting such a role.

The United States Congress

If the pessimist could not find a bright new day
of environmental reckoning in actions of the executive
branch bearing upon the Santa Barbara oil conflict, what
then of Congress? Almost certainly, he would discount
the House of Representatives, despite its role in changing the oil depletion rate in 1969 and its passing the
National Water Quality Improvement Act and the National
Environmental Protection Act in 1970. To be sure, the
Water Act *inter alia* abrogated the gross negligence requirement of the 1966 amendment to the 1924 Oil Pollution Act and substituted putting the burden of proof
upon dischargers of oil. Not only did it drastically
increase the liability of such persons to the federal
government, but also it placed considerable weight behind prompt reporting of spills. It legislated penalties of up to $10,000 and a year in jail for persons
in charge of vessels or platforms who failed to report
spills immediately to the appropriate federal agency.[18]
Yet one central fact about the House of Representatives
remained undeniable three years after the spill, none of
of the Channel bills had yet been granted a hearing by
Representative Aspinall of Colorado, Chairman of the
House Interior Committee.

At least during its 1970 session, in two ways the
Senate might have appeared a more favorable forum for
the redress of Santa Barbara grievances than either the
executive branch or the House of Representatives. First,
the thrust of general legislation about water pollution
offered by certain Democrats in the Senate appeared
much more thorough than the Republican administration's
proposals introduced by minority leader Hugh Scott.
For example, S.3471, the administration's broadest bill
on water pollution, appeared to grant discretionary
power to the Secretary of the Interior in determining
the pace and extent of the federal government's prodding
states into antipollution action.[19] Moreover S.3471
would have provided for requesting, not requiring, the
attorney general to bring suit in certain instances.[20]
By contrast, Senator Edmund Muskie's S.3687 stressed

congressional control over the secretary and envisioned a much faster pace.[21] State water pollution control boards would have had to adopt new standards, and the Secretary of the Interior would have reviewed these standards at least annually.[22] The secretary would have further been required to act where state governments seemed delinquent, and all new industrial facilities using the navigable waters of the United States as "dumping grounds" would have had to employ the best available technology as it was developed for controlling pollution.[23] Finally, it would have permitted citizen suits against those who violated the established standards.

Other Senate-originated bills also sought to press further--particularly Wisconsin Senator Proxmire's S.3181 which would have compelled the Secretaries of the Interior and of the Treasury to set by June 1971 an across-the-board scale of effluent charges for "dumping" into the navigable waters of the United States.[24] Cast in similar mode was North Dakota Senator McGovern's S.3575 empowering any individual, with the corroborating testimony of "two technically qualified persons," to bring suit against a defendant allegedly polluting air, water, or land in the United States.[25] These bills all appeared to indicate that in 1970 the Democratic-controlled Senate was taking criticisms of post-New Deal patterns of law-making much more seriously than the administration. Specifically noted was the criticism voiced by Theodore Lowi[26] that the very vagueness of post-1937 congressional delegations of power was producing "bargaining on the rule"[27] between regulatory agencies and the supposed objects of their regulation, resulting in a governmental inability to achieve a "rule of law" grounded in clear decisions about values.[28]

A second reason for suggesting that the Senate might be the most likely forum for action would of course be that bills originating in the Senate, whether terminating and compensating the leases or setting up an oil channel sanctuary, went a great deal further than the administration's bills.

One other reason lies behind the pessimist's belief that a pro-oil outcome is in the offing. That reason is the apparent unwillingness of a majority of the Senate Subcommittee on Minerals, Materials, and Fuel to report out any of the bills, at least before the end of the 92d Congress's First Session.

On the whole, however, the Senate still appeared to be the more receptive forum. The Democratic leadership might seem to be espousing the view that Congress has delegated too much authority to administrative agencies for the public interest to prevail, when the outcome is determined by the bargaining of interest groups and government agencies. But a pessimist would probably still consider it foolhardy to anticipate that the Senate committees would produce a quick correcting move.[29]

The Judiciary

Finally, what of the prospects for redress of local grievances in the third branch--the judiciary? Three years after the spill, the course of major litigation seemed unpromising. True, Union had agreed to shoulder the burden of cleanup costs on public and private beaches. Union also--without admitting liability--had agreed to settlements with a majority of the boatowners and property owners, as well as consenting to have the federal district court send claim-forms to 17,000 persons likely to have suffered damage from the slick.[30] However, against this must be weighed the broader actions asserting "ecological liability" not yet underwritten by the oil company or the government.

The litigation of three important suits--Pauley Petroleum against the government, and two parallel damage suits brought by the state, the city, and the county for half-a-billion dollars against Union and the federal government--have not proceeded far enough to warrant judgments as to their outcomes. However, neither Union Oil nor the federal government has indicated willingness to waive the respective defenses of *dammum absque injuria* and of sovereign immunity.

Moreover, as late as early 1972 the anti-oil plaintiffs in the two most interesting and potentially important cases arising from the spill had achieved no substantial success. One of these, filed originally by Santa Barbara County and City, derived its significance from the fact that the plaintiffs had later been joined by 17 defendants represented by the American Civil Liberties Union. This was the first occasion on which that redoubtable defender of the individual's civil rights had seen fit to enter the field of environmental litigation. The other case was filed by Santa Barbara District Attorney David Minier against Union for violating California's fish and game laws. Its special interest derived from Minier's seeking to hold the oil company responsible for *each day* during which the spill continued: Minier sought restitution not for one day, but for 343 days of continuing violation.

The city, the county, and the 17 individuals had undeniably received at least their share of frustration in three years of litigation. They began very expansively, by seeking a permanent injunction against Channel drilling on the grounds that the OCSL Act permitting drilling was unconstitutionally broad in its delegation of powers. By late 1969 the plaintiffs had pulled back a bit and decided to restrict their argument to the contention that the Constitution required public hearings before issuing drilling permits. In October of that year, they failed in an effort to speed matters: At that time the Supreme Court rejected their direct appeal from the federal district court. In April 1970, they fared little better: A three-judge federal circuit court in San Francisco ruled against their contention that the OCSL Act violated the Constitution on its face.[31]

In one sense, the general cause for which the plaintiffs stood was helped by the fact that on January 1, 1970 the 1969 National Environmental Policy Act went into effect. The plaintiffs sought to argue that the act, as applied to the Channel, required public hearings and statements showing that oil exploration projects would not damage the environment, before each and every permit was issued. The decision of the Interior Department to hold hearings in January 1971 on Union's and Sunoco's new drilling applications, and their ultimate denial in September 1971, rendered much of the case moot. But those very acts made the legal life of the plaintiffs more difficult: The "voluntary action" of the Interior Department dimmed the chances of creating a legally binding judicial decision. As of early 1972, the best that the county had managed was to join as "friend of the court" in a legal sequel to Interior's decision to deny those permit applications--in actions brought shortly thereafter by Union and Sunoco seeking an injunction against the federal government's order, and in the case of Sunoco, collateral damages of $200 million.

District Attorney Minier's suit was marked by drawn-out legal maneuvering over whether the federal or the state court systems should have jurisdiction. By late 1971 the suit had moved from state to federal jurisdiction, and back again to the state level. Basic to the suit was the question of whether Union's spill constituted one offense or many separate offenses. In the fall of 1971 it appeared that at last that issue would be joined and, at least initially decided. However, shortly before the November 17 date set for the hearing, both parties to the case agreed to defer argument until 1972. On January 10, 1972 the case took a turn most disagreeable to the prosecution. On that day Morton L. Barker, a retired municipal judge, pressed temporarily back into service, found the oil companies guilty on one count and dismissed the other 342. Perhaps the most curious aspect of the decision was Judge Barker's refusal to rule explicitly on the core of the issue-- whether each day's pollution was or was not a separate offense. Instead, he simply dismissed all but one count, and fined Union and her three partners $500 each. Minier thought the ruling outrageous and determined to appeal. His press statement after the decision doubt- less echoed the sentiments of many Santa Barbarans:

> It seems incredible that a retired
> judge from Los Angeles should come
> to Santa Barbara to dismiss the
> largest oil pollution prosecution
> ever filed, because in his opinion
> the oil companies had 'suffered
> sufficiently.'[32]

Even if these cases had proceeded more speedily,
it would be too early to conclude much about their ultimate disposition by the Supreme Court, their probable
final resting place. But surely, grounds for pessimism
are still substantial. There is little evidence thus
far of federal appellate willingness to embrace the
issues readily, particularly if the relief of local
wrongs entails damage awards exceeding those voluntarily
assumed by the government and the oil industry. Further,
injunctions against Channel drilling appear to contra-
dict the Supreme Court's post-1937 inclinations toward
a national solution for all wrongs.

In the best of all possible judicial worlds local
feeling should quite possibly outweigh developmental
interests. But the fact that local feelings are intense
provides no grounds for "Madisonian confidence" that a
larger political stage will improve the chances of
achieving the general long-run "good."[33] Litigation before the U.S. Supreme Court is almost certain to conflict with the fact that, for a generation, the court
has looked with disfavor on arguments asserting an unconstitutional delegation of power by Congress to another governmental branch. Neither the *Schechter Poultry
Company* case[34] nor the *Carter Coal* case[35] has been au-
thoritative since 1937. However strong arguments may
now be, in the abstract, for the court's insisting on
closer congressional scrutiny over delegations, adjudications in analogous cases seeking to boost the status
of private property rights against polluters have not
thus far been well received. Accordingly, it might be
argued that one cannot expect a receptive forum in matters of oil pollution from a court that has not resolved
the problem of aircraft noise pollution. The court is

attempting to solve the environmental problem of aircraft noise at levels that render private residential
property unlivable. For this, the court still adheres
to philosophical doctrines of trespass that justify compensation only when a plane flies directly over a property, but not when a plane producing equal noise flies
slightly outside a vertical extension of the property
boundaries.[36] If sound travels only straight up and
down, as the Supreme Court seemed to suggest in *United
States v. Causby*, then how can oil in federal waters
pollute property outside federal jurisdiction in any
legally cognizable and redressable fashion? Thus,
while the Supreme Court has been willing to treat some
issues--e.g., civil rights, criminal procedures and
reapportionment--as restricted "railroad-tickets" that
are "good for this day and train only"[37] it seems unlikely that the cause of antipollution is soon to find
so flexible a response.

PROSPECTS FOR A "PRO-SANTA BARBARA" OUTCOME

We turn now to a brief review of the most plausible
route to a "pro-Santa Barbara" outcome. Such a possibility relics heavily both on the "other side" of legislative, executive and judicial actions, and on making
two assumptions about possible changes forthcoming in
the American polity. It assumes (1) that there is a
rapidly growing concern about pollution on the part of
the American citizenry; and (2) that at least one branch
of the federal government will be able to break out of
"the charmed circle of 'Lockean' ideology and interest-
group liberalism." Both assumptions relate to the
question of systemic capacity for general, appropriate
and timely change.

Most broadly at stake is the future viability of
the western, and specifically American, perspective
about the relationship between man and earth. The perspective is grounded in the post-Renaissance belief in
man's capacity to control and to dominate his environment. Moreover, it is intimately bound up with concepts

of a "modern socioeconomic order" and of a developed polity." The following statement by a prominent American political scientist is a lucid expression of this perspective.

> Modernity begins when men develop a sense of their own competence, when they begin to think first that they understand nature and society, and then that they can control nature and society for their own purposes. Above all, modernization involves belief in the capacity of man by reasoned action, to change his physical and social environment. It means the rejection of external restraints on man, the Promethean liberation of man from control by God, faith and destiny.[38]

Classical Liberalism

Such a perspective has a set of clear economic and political derivatives in Classical Liberalism's conceptions of good politics and sound economy. Good politics creates a clear economic stage of action for the Promethean entrepreneur to exploit nature and make man free and affluent. Its American "Lockean"[39] derivatives have been perhaps the most full-blown and have found their own classical expressions in statements such as former Defense Secretary Charles Wilson's dictum that "what's good for General Motors, is good for the country." One American result, as expressed by University of Chicago economist, Milton Friedman, has been a persistent confidence that the best of all possible political and economic worlds is reached by maximizing the value-allocating role of the free-market place, and by minimizing "the use of political channels" that tend "to strain the social cohesion essential for a stable society."[40] A Lockean liberal such as Friedman, sees a great disadvantage of effecting social action through political channels:

> ...it tends to require or to enforce
> substantial conformity. The typical
> issue must be decided 'yes' or 'no';
> at most, provision can be made for
> a fairly limited number of alterna-
> tives....The number of separate
> groups that can in fact be repre-
> sented is narrowly limited, enor-
> mously so by comparison to the pro-
> portional representation of the
> market.[41]

By contrast, for Friedman, the great advantage of the market, "is that it permits unanimity without conformity; that it is a system of effectively proportional representation"[42] of competing interests and values.

Some Opposing Views

An opposing point of view concerning the Promethean capacity of man to dominate his planet has perhaps nowhere been better stated recently than in René Dubos's 1969 Pulitzer Prize-winning *So Human an Animal*.

> Biologically, man is still the great
> amateur of the animal kingdom; he is
> unique in his lack of anatomical and
> philosophical specialization. The
> range of his adaptive potentiality
> has been greatly enlarged by socio-
> cultural mechanisms that have en-
> abled him to colonize most of the
> earth. His adventuresome spirit
> now tempts him to conquer other
> worlds. But despite the success of
> launchings into space, his coloniz-
> ing days are over....

> The fact that modern man is con-
> stantly moving into new environments
> gives the impression that he is en-
> larging the range of his biological
> adaptabilities plus escaping from
> the bondage of his past. This is
> only an appearance. Wherever he
> goes and whatever he does, man is
> successful only to the extent that
> he functions in a micro-environment
> not drastically different from the
> one under which he evolved.[43]

On this showing, man is not at all Promethean in his capacity to dominate the earth; rather he is vitally dependent upon it. Further, as Athelstan Spilhaus, 1970 President of the American Association for the Advancement of Science has argued, "We must have a new industrial revolution even if a few of us have to generate it."[44] In Spilhaus' view, we have had three "unplanned" industrial revolutions so far--each of which has drawn unforeseen "disbenefits" along with its benefits.

The first industrial revolution harnessed steam and electric power but engendered air pollution. The second "multiplied things" by introducing mass production but generated the problem of solid-waste. The third was a "tremendous growth in industrial chemistry, and the ability to tailor-make chemicals in vast quantities very cheaply, for all kinds of purposes,"[45] like pesticides which have disturbed the ecological balance and poisoned waters.

According to Spilhaus, we now need a fourth "planned" industrial revolution--one which would cause us to revise our vocabulary and attitudes. It would require us to stop thinking in terms of merely using things and to begin thinking of creating "a loop back from the users to the factory which industry must close."[46] Spilhaus argues that this recycling revolution, with its concern for effluents amid affluence, could be brought about by the same industrial ingenuity

that created mass-assembly and mass-distribution. The
transformation from an exploitative "cowboy economy" to
the "spaceship economy"[47] could in this view come about
within the present economic structure.

Some economists would argue that one cannot expect
to bring about such a "revolution" within the present
economic system. They argue that the free market-place
is not capable of avoiding disastrously large social
costs from pollution because it is structured to exter-
nalize such costs, rather than to internalize them as
part of the expense of industrial corporations.[48]

We will not pass upon the merits of these two positions. However, we would suggest that, in dealing with
pollution, the political-economic structure must at
least generate a "problem-solving linkage"[49] to the
poverty sectors of the American economy.

It is quite evident that a satisfactorily rapid
shift towards a "spaceship economy" will be politically
feasible only if minority groups not now sharing fully
the American affluence can be convinced that ecology is
something more than a new device cooked up by "whitey"
to avoid the racial issue. If the American political-
economic system cannot raise the lower portion of the
curve of income inequality, there is little reason for
optimism. In this sense, pollution, population-crowding,
and racial issues are all part of one grand socioeconomic
problem. The American polity would have to break out of
its "charmed Lockean circle" at least to the extent of
generating such problem-solving linkages.

AN OPTIMISTIC PROGNOSTICATION

In the light of these general observations, what
would be an optimistic prognostication about the outcome
of Santa Barbara oil controversy? One might begin by
interpreting the course of executive actions since the
spill rather differently. That is to say, even while
admitting that the actions of the White House and

executive agencies have not brought Santa Barbara satisfaction, they indicate signs of a genuine--and eventu-
ally perhaps decisive--environmental concern. It seems
that something more than campaign rhetoric is emanating
from the Nixon Administration. One needs to balance at
least four facts against the reluctance of the White
House to follow quickly the recommendations of the first
presidential panel on oil spills and the Council of Environmental Quality.

First, drilling regulations have been tightened.
Second, the creation of the Council of Environmental
Quality and the Environmental Protection Agency suggest
more than mere lip-service to the ecological issue.
The environment was emphasized even more fundamentally,
in the provisions of 1970 Executive Reorganization Plans
Numbers 3 and 4.[50] At the very least one might be willing to give the benefit of the doubt to reorganization
plans which seem to take federal control over industrial
polluters away from those agencies so often criticized
in the past for "collusion" with industries they supposedly regulate. Third, one might look at the other
side of the administration's proposal for withdrawing
some Channel leases. True, the proposed withdrawal of
35 of the 71 leases and the creation of a partial federal sanctuary seaward of the state sanctuary did not
meet Santa Barbara's expectations. But on the other
hand, as evident in the July 22-23, 1970, hearings of
Senator Moss' Subcommittee, the oil industries were not
at all happy with even this proposal.

Their unhappiness stemmed, of course, from the
same source which engendered their disturbance over the
reduction of the oil depletion allowance from 27 percent to 22 percent. Both measures tamper with a tradition of federal behavior so carefully guarded in the
past. Once these quasi-divine rights are questioned,
how far off is the economic axe? Thus, both the administration's 1970 and 1971 bills could be interpreted as
initial executive branch actions designed to bring about
a different balance than had hitherto existed between
a great industry and a competing public interest. At a

minimum, it could be read to indicate a lessening of executive insulation from smaller pressure groups.

One might further argue for a similar interpretation of the September 1971 decision by the White House, speaking through Interior Secretary Morton, to deny Union's and Sunoco's drilling-permit applications. This involved effectively overruling the recommendations of lower echelons in the Interior Department favoring the applications. It might indeed be that political considerations rather than ecological evaluations motivated the decision. Possibly, as some cynics declared, White House anxiety about California's bloc of electoral votes in the 1972 presidential election was what lay beneath Morton's statement that the platforms would not be permitted "because of 'overriding environmental considerations.'"[51] Such an interpretation cannot have been too far from the minds of the leaders of GOO when that Santa Barbara pressure group adopted a typically American political solution to the problem: At Christmastime 1971 its leaders offered a doubtful seasonal present to the leaders of the two-party system--the suggestion that a "get-it-out" plank be put into the 1972 campaign platforms.[52]

What of other actions taken so far within Congress? One optimistic contention would note that, since the spill, Santa Barbara has managed to secure Senate hearings on the matter. Further, it is possible that some bill will be reported out in the 92nd Congress. What is likely to happen in the House of Representatives probably affords less reason for short-run optimism. But here too, the alternate interpretation might assert that the House is not likely to remain forever out of phase with the Senate.

In assessing the role of Congress at least two accomplishments should be emphasized. First, the Water Quality Improvement Act of April 3, 1970, represents a distinct shift of congressional interest-balancing away from the oil industries. Removal of the gross negligence requirement of the 1966 Amendment to the 1924 Oil

Pollution Act undeniably puts the industries in a less
favorable position for exploiting the ocean floor's oil
resources. It is likely that such a change will urge
greater caution on all oil companies. And possibly the
net result may be to make Channel operations on at least
some leases less attractive. Second, in regard to general antipollution legislation, bills such as offered
by Muskie, Proxmire, and McGovern in 1970 would have
gone far towards maximizing the pressure upon the present economic structure and its corporate components to
internalize the social costs of polluting. Congress
has not yet required such internalization of costs.
However such bills raise distinct possibilities for the
not-too-distant future. In this connection one might
note that the possibility of basing the solution to the
specific Santa Barbara problem on the action of the
California State Legislature is at least intriguing:
It is entirely possible that congressional action regarding offshore California drilling and production will
be made to depend upon state legislative action, and
that Congress may agree to match whatever restrictions
the state imposes within its own waters.[53]

In regard to the third arena for settlement of
grievances--the judiciary--it is almost certain that
the Supreme Court will become more and more receptive
to arguments raised on behalf of communities planned for
a future space-age economy and ecology. As the *Arizona
Law Review* argued four years ago, there is a constitutional peg now available, the Ninth Amendment.[54] A few
years past, it would have seemed an unlikely proposition
indeed, it was so unused. However, a 1965 landmark
decision in the field of birth control *(Griswold v. Connecticut)*[55] has suggested the potential. Thus, it could
be argued that it is much too soon to write off the Supreme Court as an active participant in Spilhaus'
"fourth revolution."

One could grant that the court has not so far exercised leadership regarding the environment in anything
like the fashion it has in regard to civil rights.
Probably the Burger Court will be much less inclined to

pioneer ahead of the rank-and-file than its predecessor
under Chief Justice Earl Warren. However, is there any
reason to believe that the Supreme Court will "lag behind?" It is doubtful that the court majority will reverse positions *vis à vis* the public no matter how much
it might reduce the gap between majority public wants
and the court's constitutional interpretations. In regard to ecology, the public seems ready for rapid movement. Congress, in the McGovern bill of 1970 allowing
antipollution suits by private citizens and in the in-
corporation of this legal provision in the Muskie antipollution bill of 1971, may in effect be moving in the
same direction.

* * * * *

To sum up, it is too early to conclude that Santa
Barbarans will in this case prevail over the awesome
power of the oil industry. Yet it is also too early to
conclude that in the foreseeable future the balance between competing values and interests will remain where
they have characteristically been since the New Deal.
The importance of the Santa Barbara oil spill and its
local aftermath lies not in its representing a last
turning point for reevaluating priorities, but rather
in its clear illumination in the public limelight of
the emerging political competition between old and new
economic and ecological lifestyles. The Santa Barbara
incident is weighted with significance for the future
functioning of the American political system.

114

to pages 1-5

1. Mobil, Texaco, and Gulf.

2. See Los Angeles *Times*, February 7, 1968, Sec. III, p. 16.

3. Pumping pressures to overcome backpressures at the kill line connection reached 400 pounds per square inch in the annulus (ring) below the blowout preventers.

4. Southeast winds are prevalent in the Santa Barbara Channel, coming from this direction 23.7 percent of the time. The spill was southeast of Santa Barbara.

5. See the annual review of the 100 largest California corporations for an erroneous statement to the same effect in the Business and Financial Section of the Los Angeles *Times*, May 11, 1969, Sec. H, p. 2. On June 16, 1969, *Advertising Age* took the Santa Barbara Chamber of Commerce to task for engaging in misleading advertising, which Union had helped pay for--reported in Santa Barbara *News-Press*, June 19, 1969, Sec. A, p. 1. Such advertising had already been roundly denounced as shortsighted by local citizens, conservationists, the local newspaper and particularly by the "Get Oil Out" Committee (GOO), led by a former Democratic state senator, Alvin Weingand. GOO had gathered more than 100,000 signatures calling for the cessation of all Channel oil production. See the Santa Barbara *News-Press*, May 6, 1969, Sec. A, p. 1, and June 8, 1969, Sec. A, pp. 1, 4.

6. The suit was also filed against her partners on the platform, Mobil, Texaco, Gulf, and against Peter Bawden Drilling Inc., the drilling contractor. Hereafter, unless otherwise noted, "Union" is understood to include these other corporate defendants.

NOTES
to page 5

7. That is, Santa Barbarans who did not own boats
or shoreline property, but who were accustomed to swimming, surfing or beach-strolling.

8. A confidential well-placed source told the
authors unofficially that the total claims for all suits
against Union amounted to some $8 billion, but this
seems excessive.

9. See Gene T. Kinney, "Offshore Sales Face Uncertain Future," *Oil and Gas Journal* 67(9):84-86 (March 3,
1969).

10. Pauley Petroleum v. the United States, dock-
eted as No. 197-69, U.S. Court of Claims.

11. Under *California Penal Code*, section 373(a).
Minier had barely taken the first step in notifying the
defendants when, pursuant to a request in federal court
by Humble Oil Company and others, the judge of the U.S.
District Court for the Central District of California
restrained Minier on the grounds that he was exceeding
his jurisdiction. See Chapter VII, note 52, for further
discussion of this suit.

12. County of Santa Barbara and City of Santa
Barbara v. Walter J. Hickel, No. 69-636-ARH (U.S. District Court for the Central District of California,
filed April 4, 1969). Other named defendants were D.W.
Solanas, Union Oil, Pauley Petroleum, Mobil Oil, Phillips Petroleum, Western Offshore Drilling and Exploration Company, Standard Oil of California, Humble Oil,
Peter Bawden Drilling, Inc., Gulf Oil, Texaco, and
Atlantic-Richfield Oil.

13. 43 *United States Code* (U.S.C.), Sec. 1331-
1343 (1964).

NOTES
to pages 5-7

14. For a recent argument urging serious reconsideration of the "unconstitutional delegation of powers" doctrine, see Theodore J. Lowi, *The End of Liberalism* (New York: Norton, 1969).

15. To borrow an expression attributed to Eugene W. Standley, staff engineer in Interior Secretary Hickel's office. In a memorandum from Standley to J. Cordell Moore, Assistant Secretary of the Interior, dated February 15, 1968, Standley is quoted as saying that the department had decided not to hold hearings on the federal oil leasing proposals in the Santa Barbara Channel because "we preferred not to stir up the natives ...." Santa Barbara *News-Press*, March 11, 1969, Sec. A, p. 1.

16. Rose v. Socony-Vacuum Corp., 54 *Rhode Island Reports* (R.I.) 411; 173 *Atlantic Reporter* (A.) 627 (1934).

17. Hence our frequent reliance in this essay upon the *News-Press* for factual data, and particularly upon the excellent reporting and commentaries by *News-Press* reporter, Robert E. Sollen.

18. See Senate bill, S. 1219, 91st Cong., 1st Sess. (1969), which sought to delay Channel drilling until the Secretary of the Interior had completed a study of drilling, production and transportation of oil on outer continental shelf leases, and which sought methods of phasing out Channel oil leases. Its amended 1970 version is discussed in Chapter VI.

19. H.R. 7074, 91st Cong., 1st Sess. (1969), would prohibit all future Channel leases and rescind those in effect. H.R. 3122, 91st Cong., 1st Sess. (1969) would amend the Federal Water Pollution Control Act, 33 U.S.C. Sec. 466ff. (1964) by imposing strict liability for

NOTES
to pages 7-8

discharges of oil into or upon the navigable waters and adjoining coastline of the United States. H.R. 3120, 91st Cong., 1st Sess. (1969), would bar the Secretary of the Interior from permitting any mineral exploration or development within a two-mile federal "buffer zone" off the Santa Barbara Channel State Sanctuary. See Chapter VI.

20. According to the 1968 provisional census, the population of Santa Barbara County was 254,900, while that of Ventura County was 351,000. State of California, Dept. of Finance, *California Statistical Abstract* (1968) p. 12.

21. Teague estimated that his Ventura County constituents were approximately equally divided. Santa Barbara *News-Press*, February 19, 1969, Sec. A. p. 4. The City Council and Chamber of Commerce of that county's largest city, Ventura, expressly declined to go along with the Santa Barbara City Council's call for abandonment of further drilling. Santa Barbara *News-Press*, March 7, 1969, Sec. C, p. 12. Santa Barbara received speedier and greater "aid and comfort" from more distant littoral counties, particularly from Santa Cruz, whose Board of Supervisors passed on February 25, 1969, resolution #102-69 declaring "opposition to oil exploration and drilling operations which could create...oil spillage in Monterey Bay, the Santa Barbara Channel, or in other coastline areas of the State of California." Telephone communication between the Clerk of the Board of Supervisors, Santa Cruz County, California, and Phil G. Olsen on March 31, 1971.

22. Quoted by Robert Sollen, "Oil Company President Thinks Firms Might Favor Pullout," Santa Barbara *News-Press*, March 7, 1969, Sec. A, p. 1.

NOTES
to pages 8-9

23. "Channel Oil Drillers Tapping Little But Trouble," Santa Barbara *News-Press*, March 9, 1969, Sec. A, pp. 18-19. Reprinted from the *Wall Street Journal*.

24. Santa Barbara *News-Press*, March 10, 1969, Sec. A, p. 1.

25. Santa Barbara *News-Press*, June 22, 1969, Sec. A, p. 1.

26. Santa Barbara *News-Press*, June 10, 1969, Sec. B, p. 1.

27. See particularly the prophetic statements of Fred Eissler, then a national director of the Sierra Club on November 20, 1967 to the effect that he considered it incredible that the South Coast communities, with everything to lose from a major oil spill, had not been asked to participate in consultation with the various commercial interests and federal agencies. U.S. Army Engineer District, Los Angeles Corps of Engineers, *Public Hearings on the Construction of Platform Hogan in the Santa Barbara Channel Held at Santa Barbara, California, November 20, 1967*, pp. 59-71.

28. For an indication that the Army Corps of Engineers considered esthetic and economic implications to be beyond its purview, see Chapter V.

29. The sanctuary was, and remains, 16 miles in length. See the Shell-Cunningham Act, *California Public Resources Code Annotated* (Cal. Pub. Res. C.A.), Sec. 6801ff. (1963). A 1957 amendment to the Shell-Cunningham Act provided for abrogation of the sanctuary in the event that future federal oil leasing proved to be depleting the state-owned reserve within the sanctuary. Ibid., Sec. 6872.1.

NOTES
to pages 9-11

30. Under the direction of David Bickmore, then
Santa Barbara County Oil Well Inspector. See, Oil Well
Inspection Department, Santa Barbara County, California,
*Offshore-Onshore Petroleum Study, Santa Barbara Channel,
Phase I. Effects of Federal Leasing Outer Continental
Shelf (Preliminary Report).* It was inevitably a "rush
job." Throughout this study, the publication will be
referred to as *Offshore-Onshore Petroleum Study.*

31. As of July, 1970, the camouflaging was not
markedly effective--notwithstanding a somewhat tardy
order given in Department of the Interior, Geological
Survey, Conservation Division of Oil and Gas Operations
West Coast Region, Outer Continental Shelf Order No. 9,
*Notice to Lessees and Operators of Federal Leases in
the Outer Continental Shelf, West Coast Region, Special
Requirements for Fixed Platforms in the Santa Barbara
Channel Area, January 3, 1968,* requiring that each platform be "camouflaged by paint" and that "Each platform
receiving approval will be subject to a design which
will give the best appearance available to the industry."
On a clear day, from any unobstructed elevation above
50 feet, oil platforms up to nine miles offshore dominate the seascape. This is not surprising since visi-
bility conditions frequently permit the beholder on a
peak in Santa Barbara to distinguish sunny and shady
spots on Santa Cruz Island cliffs some 25 miles out to
sea.

32. See Gabriel A. Almond and James S. Coleman,
eds., *The Politics of Developing Areas* (Princeton, New
Jersey: Princeton University Press, 1960), p. 16, for
a discussion of the theoretical concept of articulating
and aggregating competing political interests.

33. Herbert Wechsler, "Toward Neutral Principles
of Constitutional Law," 73 *Harvard Law Review* 1:1-35
(1959-1960).

NOTES
to page 11

34. See Theodore J. Lowi, *The End of Liberalism*
(New York: Norton, 1969), particularly pp. 55–100.

35. Ibid., particularly pp. 147–153.

NOTES
to pages 13-14

1. *Offshore-Onshore Petroleum Study*, Sec. 2, p.
26. See Sec. 1, pp. 1-3 and Addendum to Sec. 1, pp.
7-18, for an historical narrative of oil development
in Santa Barbara County.

2. One well, Lutton Bell #1, set a world record
by producing 2,400 barrels per day and thus paying for
itself in the first two weeks.

122

NOTES
to pages 19-20

1. Use of *Torrey Canyon* data is particularly dubious in view of the apparently deleterious effects upon wildlife by chemical dispersants used.

2. If, however, one wishes to lump together the two essentially different economies of the South Coast (which contains a majority of the population) and of the "North County" mixed farming and industrial areas, oil's contribution is considerably greater: 9.62 percent of total county revenues. *Offshore-Onshore Petroleum Study*, Sec. 5, p. 1.

3. Santa Barbara County, California, Planning Department, *General Plan: Basis for Planning, Santa Barbara County General Plan Studies 1962*, p. 7.

4. Santa Barbara *News-Press*, May 1, 1967, Sec. A, p. 1.

5. The vote might have been more heavily weighted against the oil company had the issue been stated more clearly on the ballot. The ballot read, "Shall Santa Barbara County Zoning Ordinance No. 1913 which amends Ordinance No. 661 to permit an oil and gas processing facility to be located east of the City of Carpinteria, California, be adopted?" Many voters were of the opinion that a "yes" vote constituted a vote against the aspirations of the oil company when the opposite was true.

6. The Federal Water Pollution Control Administration reported that they had received 3,000 communications. The White House reported that 1,300 letters had been received. Congressional offices estimated that they had received 7,750. Santa Barbara *News-Press*, February 19, 1969, Sec. A, p. 4.

NOTES
to pages 20-22

7.   Santa Barbara *News-Press*, February 19, 1969,
Sec. A, p. 4.

8.   Santa Barbara *News-Press*, January 11, 1970,
Sec. A, p. 8.

9.   Santa Barbara *News-Press*, March 1, 1970, Sec.
B, p. 1.

10.   Santa Barbara *News-Press*, March 3, 1970, Sec.
B, p. 1.

11.   Santa Barbara *News-Press*, May 10, 1970, Sec.
B, p. 3.

12.   See the admission by the United State Geological Survey spokesman that the early USGS estimate of
500 barrels per day was too low by a factor of two.
Santa Barbara *News-Press*, June 21, 1970, Sec. B, p. 3.

13.   In making these estimates, Allen has applied
sophisticated scaling techniques to aerial photos of
the spill.

14.   The USGS figures were lower--420 to 2,100 gallons per day.

15.   Walter J. Mead and Philip E. Sorenson, "The
Economic Cost of the Santa Barbara Oil Spill." Unpub-
lished paper prepared for the Santa Barbara Oil Symposium; University of California, Santa Barbara, December
16-18, 1970, p. 3.

16.   Particularly from late October to early November, and again in December, 1969. In the latter case,
however, the cause was another accident on Union's Platform A. See note 3 of this chapter.

NOTES
to pages 24-27

17. While the *Torrey Canyon* was not directly owned
by Union, it was transporting oil destined for Union's
refineries when it ran aground.

18. Statement by Professor J.H. Connel at Public
Symposium on Oil Pollution at the University of California, Santa Barbara, April 12, 1969. This compares
favorably to 450 that survived out of 5,811 living birds
that were brought in for treatment during the *Torrey
Canyon* disaster.

19. Jack E. Hemphill, California Bureau of Sport
Fisheries and Wildlife at *Hearing on S.* 7 before the
Subcommittee on Air and Water Pollution of the Senate
Public Works Committee, 91st Cong., 1st Sess., at Santa
Barbara, California, on February 24-25, 1969.

20. Santa Barbara *News-Press*, January 9, 1970,
Sec. B, p. 12

21. David Snell, "Iridescent Smell of Death,"
Life 66(23):22-27 (June 13, 1969). The Department of
the Interior has stated that there is no evidence that
oil pollution from the five-month-old Union leak killed
any seals or sea lions on San Miguel. Miss Jodi Bennett,
scientist for the University of California, Santa Barbara, museum of zoology, made an independent study of
the effects of the oil on these animals. She is of the
opinion that the oil has killed a number of seals and
sea lions and that there has been a serious impact on
the breeding cycle of the animals. For a comparision
of the Interior and Miss Bennett's opposing views, see
Santa Barbara *News-Press*, June 29, 1969, Sec. A, p. 1.

22. "Oil's Aftermath," *Time* 97(9):37 (March 1,
1971).

23. Loc. cit.

NOTES
to pages 27-31

24. Max Blumer, Howard L. Sanders, J. Fred Grassle and George R. Hampson, "A Small Oil Spill," *Environment* 13(2):3 (1971).

25. Santa Barbara *News-Press*, February 19, 1969, Sec. D, p. 9.

26. Santa Barbara *News-Press*, June 29, 1970, Sec. A, p. 1.

27. Santa Barbara *News-Press*, June 22, 1969, Sec. A, p. 3.

28. Santa Barbara *News-Press*, October 23, 1969, Sec. D, p. 8.

29. See note 15, above.

30. Emphasis added. It is the last phrase of this quotation and the final sentence of the following quotation which merit notation.

31. The reference is to the *Torrey Canyon*, which was carrying Union oil when it went aground off the Cornish coast.

32. Santa Barbara *News-Press*, December 21, 1969, Sec. A, p. 1; December 22, 1969, Sec. A, p. 1; and December 25, 1969, Sec. A, p. 4.

33. Santa Barbara *News-Press*, December 31, 1969, Sec. A, p. 1. It should be noted that Hartley was speaking of the third platform on Tract 402. The second was already under construction.

34. Santa Barbara *News-Press*, November 10, 1971, Sec. F, p. 9.

NOTES
to pages 32-36

1. Santa Barbara *News-Press*, November 25, 1969, Sec. A, p. 1.

2. Statement by Robert O. Anderson, Chairman of the Board of Atlantic-Richfield, addressing the 13th National Conference of the U.S. Commission for UNESCO in November 1969.

3. This view has been expressed by Frank Stead, who is presently a consultant in Piedmont, California and former Chief of the Division of Environmental Sanitation, Department of Public Health, State of California.

4. See statement to this effect by former Secretary of the Interior Stewart Udall, Santa Barbara *News-Press*, February 7, 1968, Sec. A, p. 11.

5. Evidence presented by Walter J. Mead, professor of economics at the University of California, "Economics of Oil Has Many Facets," Santa Barbara *News Press*, February 9, 1969, Sec. A, pp. 14, 15.

6. Santa Barbara *News-Press*, December 18, 1969, Sec. A, p. 5.

7. Santa Barbara *News-Press*, February 20, 1970, Sec. A, p. 3.

8. "Finding Lemonade in Santa Barbara's Oil," *Saturday Review* 52(19):18-21 (May 10, 1969).

9. This argument as to national interest is by no means so "far out" as it may seem. See for instance the concern manifested by the former Secretary of Defense Robert S. McNamara in address to the University of Notre Dame, May 1, 1969. "The Excessive Population Growth," *Vital Speeches* 35(16):500-505 (June 1, 1969).

NOTES
to page 36

10.   See Kenneth Boulding, "The Economics of the
Coming Spaceship Earth," in Garrett de Bell, ed., *The
Environmental Handbook* (New York: Ballantine, 1970),
pp. 96-101. By "cowboy economy" Boulding means the
"open economy" possible in the past when the earth's
natural resources and environmental resiliency seemed
well-nigh infinite--"the cowboy being symbolic of the
illimitable plains and also associated with reckless,
exploitative, romantic, and violent behavior, which is
characteristic of open societies." To this Boulding
contrasts the "spaceman economy" of the future, wherein
"the earth has become a single spaceship, without unlimited reservoirs of anything, either for extraction
or for pollution, and in which, therefore, man must
find his place in a cyclical ecological system which is
capable of continuous reproduction of material form
even though it cannot escape having inputs of energy."
For a discussion of economic difficulties of a market
economy involved in the transition from "cowboy" to
"spaceship" economy, see M. Mason Gaffney, "Welfare Economics and the Environment," ibid., pp. 88-101. For a
more pessimistic discussion, see Karl William Kapp,
Social Costs of Business Enterprise (New York: Asia Publishing House, 1963). For a stimulating analysis of
possible means whereby the public sector could recalcu-
late its costs and benefits analysis in making public
investment decisions, see Arthur Maass, "Benefit-Cost
Analysis: Its Relevance to Public Investment Decisions,"
Quarterly Journal of Economics 80(2):208-226 (1966).
For discussion of the cowboy-spaceship dichotomy's bearing upon contemporary political science, see A.E. Keir
Nash, "Pollution, Population, and the Cowboy Economy,"
Journal of Comparative Administration, 2(1):109-128
(1970) particularly pp. 115-126.

11.   Santa Barbara *News-Press*, October 30, 1969,
Sec. A, p. 11.

128

NOTES
to pages 36-39

12.   See interviews with Woods Hole scientists in
Santa Barbara *News-Press*, March 11, 1970, Sec. A, p. 1.

13.   "The Real Meaning of Alaskan Oil Finds," *U.S.
News and World Report* 66(9):66-67 (March 3, 1969). Oil
production in the frozen north seems preferable to
Santa Barbarans although such exploitation might offend
the more ardent conservationist who would defend both
Santa Barbara and Alaska.   Indeed, conservationists consider the Alaskan tundra extremely fragile and suscep-
tible to pollution.   See George Laycock, "Whittling
Alaska Down to Size," *Audubon* 71(3):66-87 (1969), for a
discussion on oil production in Alaska.

14.   J. Ryan and S. Wells, *Regional Economic Impact
of U.S. Oil Shale Industry, Public Policy Studies of a
United States Oil Shale Industry, Number 1*   (Denver:
Denver Research Institute, University of Denver, 1966)
p. 11.

15.   Geological Survey Circular 523, D. Duncan and
V. Swanson, *Organic-Rich Shale of the United States and
World Land Areas 1965*, p. 9.

16.   U.S. Department of the Interior, Bureau of
Mines, *Commodity Statements*, January 1968, reprinted in
Congressional Record, April 1, 1968, p. S3718.   For two
possible solutions--*in situ* extraction and growing vege-
tation over spent shale after a few years' weathering--
see Note, "Problems and Politics of Oil Shale Development," 19 *Stanford Law Review* 1:190-216 (1966).

17.   "Letter," Santa Barbara *News-Press*, June 8,
1969, Sec. E, p. 8.

18.   "What Would Earthquakes Do To Offshore Wells?,"
Santa Barbara *News-Press*, March 2, 1969, Sec. A, pp. 1,
12.

NOTES
to pages 39-42

19. Most geologists place the beginning of the
Tertiary Period somewhere between 60 and 70 million
years ago and that of the Quaternary Period between 1
and 2 million years ago.

20. The Oligocene, Miocene, and Pliocene epochs
are subdivisions of the Tertiary period. The Oligocene
epoch started approximately 40 million years ago, was
followed by the Miocene epoch ending about 11 million
years ago, and by the Pliocene epoch ending about 2
million years ago. See Thomas F. Dibblee, Jr., *Geology
of the Central Santa Ynez Mountains, Santa Barbara
County, California,* Bulletin 186 (Sacramento: California Division of Mines and Geology, 1966); and Geologic
Survey Professional Paper 679, *Geology, Petroleum Development, and Seismicity of the Santa Barbara Channel Region, California* (Washington, D.C., 1969) for more complete discussions of the geology of the Santa Barbara
Channel.

21. Santa Barbara *News-Press*, May 12, 1967, Sec.
B, p. 1.

22. For an official description of events that
occurred when the well blew out see U.S. Department of
the Interior, Geologic Survey, *First Field Report on
Union Company Lease OCS P-0241, Platform A, Well #21,*
February 4, 1969.

23. In testimony before Senate investigators, Fred
Hartley, President of Union Oil, was quoted to the effect that the blowout would probably not have occurred
if the hole on Well A-21 had been cased to 2,000 feet.
However, he claimed that this was "20-20 hindsight" and
that no one in the industry or the Department of the
Interior had guessed beforehand that drilling practices
and regulations were inadequate for the situation.
Quoted by Gene Kinney, "Stricter Offshore Leasing, Pollution Rules Likely," *Oil and Gas Journal* 67(6): 43 (February 10, 1969).

NOTES
to pages 42-44

24. For the sake of uniformity, throughout this essay federal government terminology is used in describing drilling and safety procedures. A "string" is the term used in the industry to describe either the pipe attached to the drill or the total length of the various casings. Normally, a drill string is constructed of 90-foot "stands," which are in turn composed of three 30-foot "sections" of pipe.

25. For a more complete comparison of California and federal regulations for casing requirements on offshore wells see the statement signed by W. Bailey, Chief Deputy Supervisor, California Division of Oil and Gas, *Requirements for Offshore Wells Drilled from Ships or Platforms*, October 2, 1968, and order signed by D. W. Solanas, Regional Oil and Gas Supervisor, West Coast Region, Department of the Interior, Geologic Survey, Conservation Division of Branch of Oil and Gas Operations West Coast Region, *Outer Continental Shelf Order No. 2, Notices to Lessees and Operators of Federal Leases in the Outer Continental Shelf, West Coast Region*, March 31, 1965.

26. Answer to question posed at the Public Symposium on Oil Pollution, University of California, Santa Barbara, April 12, 1969. But see also Memorandum from Department of the Interior Regional Staff Engineer, M. V. Adams, to the Department of the Interior Oil and Gas Supervisor, Pacific Region, in which Mr. Adams quoted a California official to the effect that California followed essentially a program of two strings of casing, "but there were variances to suit particular conditions." Memorandum dated February 6, 1969.

27. See note 22 above, *First Field Report*.

28. Santa Barbara *News-Press*, March 30, 1969, Sec. A, p. 3.

NOTES
to pages 45-48

29.   Thomas H. Gaines, Coordinator, Air and Water
Conservation, Union Oil Company of California, at *Hear-
ings on S. 7* before the Subcommittee on Air and Water
Pollution of the Senate Public Works Committee, 91st
Cong., 1st Sess., at Santa Barbara, California on Feb-
ruary 24-25, 1969.

30.   Santa Barbara *News-Press*, February 5, 1970,
Sec. B, p. 6.

31.   Santa Barbara *News-Press*, June 21, 1970, Sec.
B, p. 3.

32.   Ibid., p. 12.

33.   "Insurance and the Environment," *The Journal
of Insurance* 31(4):13-20 (1970), p. 16.

34.   Ibid., p. 17.

35.   *Oil Insurance Limited* (an information book-
let published by Oil Insurance Limited), December 2,
1970, p. 1 of "Shareholders' Agreement."

NOTES
to pages 49-52

1. For a discussion of the "marble cake" concept of federalism see Morton Grodzins, *The American System* (Chicago: Rand-McNally & Co., 1966), Part II, pp. 60-80.

2. For a general discussion of federalism see Daniel J. Elazar, *American Federalism: A View from the States* (New York: Thomas Y. Crowell Co., 1966).

3. The Federalist No. 10 in Alexander Hamilton, James Madison, John Jay, *The Federalist* (New York: Charles Scribner's Sons, 1921).

4. Grant McConnell, *Private Power and American Democracy* (New York: Alfred A. Knopf, 1951), p. 190.

5. William H. Riker, *Federalism: Origin, Operation, Significance* (Boston: Little, Brown and Co., 1964), p. 154.

6. 43 U.S.C. Secs. 1331-1343 (1964). See Regulations Pertaining to Mineral Leasing, Operations, and Pipelines on the Outer Continental Shelf as contained in Title 30 and Title 43 *Code of Federal Regulations* (C.F.R.) and the Outer Continental Shelf Lands Act (67 *Statutes at Large* [Stat.] 462) United States Department of the Interior, Bureau of Land Management, April 1968.

7. The federal government had exercised this authority on the basis of executive orders issued by President Truman in 1945, with the Secretary of the Navy acting as his agent. Executive Order No. 9633, 3 C.F.R. (1943-1948), p. 437; Exec. Proclamation No. 2667, 3 C.F.R. (1943-1948), p. 47.

8. Oil and Gas and Sulphur Operations in the Outer Continental Shelf, 30 C.F.R., Sec. 250 (1969), p. 12.

NOTES
to pages 52-54

9. Ibid., p. 42.

10. It was repealed on April 3, 1970, by the passage of the Water Quality Improvement Act of 1970, discussed below in Chapters VI and VIII (*Public Law* [P.L.] 91-224, 84 Stat. 91).

11. By the same act, P.L. 91-224. For the sake of clarity throughout this essay we use the earlier term in discussions about events before April 3, 1970.

12. 33 U.S.C. Sec. 466 (1966 supp. IV).

13. Under the U.S. President, *National Multi-Agency Oil and Hazardous Materials Pollution Contingency Plan*, the On-Scene Commander (The Commanding Officer of the Santa Barbara Coast Guard Group in the case of the Channel spill) has the authority to authorize the use of dispersants and other chemicals. See pages 8 and 9 of the cited plan. However, Annex XIII of the plan outlines the Department of the Interior, Federal Water Pollution Control Administration Policy On the Use of Chemicals to Treat Floating Oils. By personal communication, the On-Scene Commander of the Santa Barbara spill, Lt. George H. Brown III, U.S. Coast Guard, informed the authors that he relied very heavily on advice by the FWPCA before authorizing any use of chemical dispersants by Union.

14. Section 10, Rivers and Harbors Act of 1899, 30 Stat. Sec. 1151 (1897-1899) 33 U.S.C. Sec. 403 (1964); 67 Stat. Sec. 463 (1953).

15. 33 U.S.C. Sec. 407 (1964). Note that the 1899 Act does not prohibit street or sewage discharges.

16. Statement of Col. Norman E. Pehrson, District Engineer, U.S. Army Corps of Engineers at *Hearings on S.7* before the Subcommittee on Air and Water Pollution of the Senate Public Works Committee, 91st Cong., 1st Sess.,

(1969), Santa Barbara, California, February 21, 1969,
p. 6. See also the corps' statement of January 3, 1970,
Santa Barbara *News-Press*, January 14, 1970, Sec. C, p.
1.

17. *National Multi-Agency Oil and Hazardous Mate-
rials Pollution Contingency Plan*, September 1968, in
Oil Pollution, U.S. Congress, House of Representatives,
Committee on Merchant Marine and Fisheries, Hearings,
91st Congress, 1st Session, on H.R. 6495, H.R. 6609,
H.R. 6794, and H.R. 7325. Serial No. 91-4 (Washington,
D.C.: 1969).

18. Cal. Pub. Res. C.A. Sec. 6871.3 (Deering 1963).

19. Ibid., Sec. 6873(b).

20. Ibid., Sec. 6873.2.

21. *California Water Code Annotated* (Cal.Wat.C.A.)
Sec. 13005 (Deering 1964). For a general discussion of
the development of pollution laws in California, see
Adolphus Moskovitz, "Quality Control and Re-use of Water
in California," 45 *California Law Review* 5:586-603
(1957).

22. *California Fish and Game Code Annotated* (Cal.
F.& G.C.A.) Sec. 5650 (Deering 1957).

23. *California Harbors and Navigation Code Annotated* (Cal. Harb. & Nav. C.A.) Sec. 151 (Deering 1964).

24. Paul De Falco, Jr., Regional Director of the
Pacific Southwest Region of the Federal Water Pollution
Control Administration, Department of the Interior, at
Hearing on S. 7 before the Subcommittee on Air and Water
Pollution of the Senate Public Works Committee, 91st
Cong., 1st Sess., at Santa Barbara, California, on
February 24-25, 1969, p. 6.

NOTES
to pages 59-63

1. Presumably based upon the Fifth Amendment.

2. See Chapter I, note 10.

3. See Chapter IV, note 25.

4. 30 *Coded Federal Regulations* (C.F.R.) (May 1969) Sec. 250.12(c).

5. Ibid., Sec. 250.40(a).

6. Ibid., Sec. 250.43(a).

7. Ibid., Sec. 250.41(a).

8. *Hearings Before the Senate Subcommittee on Minerals, Materials, and Fuels*, March 13-14, 1970 (Washington, D.C.: 1970), p. 3.

9. Los Angeles *Times*, June 2, 1969, Sec. I, p. 1.

10. See note 8 above.

11. Santa Barbara *News-Press*, June 3, 1969, pp. A1, A5, and C10. See also, Santa Barbara *News-Press*, June 5, 1969, Sec. A, p. 1.

12. Santa Barbara *News-Press*, April 28, 1970, Sec. B, p. 1.

13. Humble Oil turned back Tract 395 which it had purchased for $45,262,080. See Santa Barbara *News-Press*, February 10, 1970, Sec. A, p. 1. Signal Oil returned a lease which had cost it much less--$225,558. Signal had never even drilled on it. Santa Barbara *News-Press*, April 28, 1970, Sec. A, p. 1.

NOTES
to pages 64-69

14. Santa Barbara *News-Press*, November 26, 1969,
Sec. A, p. 1.

15. "Next Move on Oil Imports: Tariffs to Replace
Quotas," *Chemical Week* 106(1):16-17 (1970).

16. New York *Times*, July 26, 1970, Sec. F, p. 1.

17. 84 Stat. 91.

18. Compare P.L. 91-224 Title I, Sec. 108, with
H.R. 3122, 91st Cong., 1st Sess. (1969).

19. Public Law Water Quality Improvement Act,
April 3, 1970. 91-224 Title I, Sec. 11(f).

20. Loc. cit.

21. Ibid., Sec. 11(b).

22. Ibid., Sec. 11(o).

23. S. 3575, 91st Cong., 2d Sess. (1970), Sec.
3(a).

24. See Chapter I, notes 18 and 19.

25. S. 4017, 91st Cong., 2d Sess. (1970).

26. H.R. 12541, 91st Cong., 1st Sess. (1969-70).

27. H.R. 18159, 91st Cong., 2d Sess. (1970).

28. Testimony at the July 21-22, 1970 (91st Cong.,
2d Sess.) Hearings before the Subcommittee on Minerals,
Materials, and Fuels of the Committee on Interior and
Insular Affairs, United States Senate, on S. 1219,
S. 2516, S. 3351, S. 4017, S. 3516, and S. 3093. At-
tended by A.E. Keir Nash, held in Washington D.C.

NOTES
to pages 69-75

29. Santa Barbara *News-Press*, October 20, 1969,
Sec. A, p. 1.

30. H.R. 19548, 91st Cong., 2d Sess.

31. Ibid., Sec. 6.

32. Oral statements on July 22, 1970, morning
session. See note 28 above.

33. Senator Alan Cranston's written statement
(also delivered orally) before the subcommittee, July
22, 1970, Sec. B, p. 2. See note 28 above.

34. Loc. cit.

35. Ibid., Sec. B, p. 3.

36. Ibid., Sec. A, p. 2.

37. Ibid., Sec. B, p. 1.

38. Loc. cit.

39. Santa Barbara *News Press*, February 4, 1970,
Sec. B, p. 1.

40. Santa Barbara *News-Press*, June 26, 1970, Sec.
B, p. 1.

41. Santa Barbara *News-Press*, February 11, 1970,
Sec. A, p. 4.

42. *Hearings before the Senate Subcommittee on Minerals, Materials, and Fuels*, March 13-14, 1970 (Washington D.C., 1970), p. 14.

43. Loc. cit.

NOTES
to pages 75-78

44. Ibid., p. 15.

45. Santa Barbara *News-Press*, February 11, 1970, Sec. A, p. 1.

46. Santa Barbara *News-Press*, January 9, 1970, Sec. A, p. 1.

47. Santa Barbara *News-Press*, December 28, 1971, Sec. A, p. 9.

NOTES
to pages 79-81

1. Santa Barbara *News-Press*, October 22, 1969, Sec. A, p. 8.

2. *Marbury v. Madison*, 1 Cranch 137 (1803).

3. Ibid., p. 163.

4. Loc. cit.

5. See Chapter I, note 16.

6. Undeniably, as a matter of actual legal history--though not as a common *post hoc* judicial interpretation thereof--what follows is an oversimplification. See, for example, the attitudinal complexities of so influential an early Justice as Chancellor Kent discussed in Joseph Dorfman, "Chancellor Kent," 61 *Columbia Law Review* 7:1290-1317 (1961).

7. And that preferably on a day when the "ultranationalist" Justice McLean chanced to be absent.

8. See, for instance, Charles River Bridge v. Warren Bridge, 11 *Peters* (Pet.) 420, 9 *Lawyers' Edition, United States Supreme Court Reports* (L. Ed.) 773 (1837).

9. See E.F. Roberts, "Negligence: Blackstone to Shaw to?, An Intellectual Escapade in a Tory Vein," 50 *Cornell Law Quarterly* 2:191-216 (1964-1965).

10. Walt W. Rostow, *The Stages of Economic Growth: A Non-Communist Manifesto* (Cambridge: The University Press, 1960), p. 38.

11. Estimated--1860-1920. Ibid., p. 11.

12. The phrase is the late Vernon L. Parrington's. See his *Main Currents in American Thought* (New York: Harcourt, Brace, 1930).

NOTES
to pages 82-84

13.   Fletcher v. Rylands, Exchequer Chamber 1886
Law Reports (L.R.) 1 Ex. 265.   Rylands and Horrocks v.
Fletcher, House of Lords 1868 L.R.   3 H.L. 330.

14.   Lord Cairns added the "when clause" on appeal
to the House of Lords.   The lower court had supposed
that the maxim also applied to activities which were
not extrahazardous.

15.   See John D. Knodell, "Liability for Pollution
of Surface and Underground Waters," 12 *Rocky Mountain
Mining Law Review* 33 (1967).

16.   Pennsylvania Coal Co. v. Sanderson, 113 *Pennsylvania State Reports* (Pa.) 126; 6 *Atlantic Reporter*
(A.) 453 (1886).

17.   Ibid., 6 A. 456.

18.   86 Pa. 401; and 94 Pa. 302.

19.   113 Pa. 126; 6 A. 465.

20.   See the reasoning, ibid., pp. 459-463.   Hardly
a model of clarity.

21.   See similarly, ibid., p. 464.

22.   See ibid., p. 455.

23.   Ibid., p. 464--which fact naturally made the
merits of the coal company's claim to "natural use" the
more natural.

24.   Ibid., p. 459.

25.   The year when the U.S. Supreme Court decided
the crucial case of Wabash, St. Louis & Pacific Railway

NOTES
to page 84

Company v. Illinois, 118 *United States Supreme Court
Reports* (U.S.) 447 (1886). A six to three Court majority held unconstitutional the Illinois legislature's
attempt to prevent railroads from charging greater rates
for transporting freight from points of origin where
they had an effective monopoly than from places where
they were engaged in competition with each other. In
the Wabash case, the railroad was charging 66 percent
more for shipping from one town in Illinois to New York
than for shipping from another Illinois town to the same
destination--despite the fact that in the cheaper case
the distance to New York was 86 miles less. Notwith-
standing this practice, the Supreme Court held that Illinois' attempt to regulate rates contravened the commerce
clause. The result was that, a year later, Congress established the Interstate Commerce Commission.

26. See Eakin v. Raub, 12 *Sergeant and Rawle*
(Pennsylvania State Supreme Court) (Pa.S.Ct.) 330 (1825),
for Gibson's opinion--perhaps the most cogent judicial
attack upon judicial supremacy in that era. See Charles
Warren, *The Supreme Court in United States History* (Boston: Little, Brown, and Company, 1926), pp. 711-713,
for an account of why this distinguished Pennsylvania
jurist was never raised to the federal Supreme Court.

27. See Woodrow Wilson, *Congressional Government:
A Study in American Politics* (Boston: Houghton, Mifflin
Co., 1885). The structures of both rule making and rule
application made it more difficult for the "Jeffersonian
individual" to hold in account the "Hamiltonian enterprise."

28. See Justice Oliver Wendell Holmes's eloquent
protest against this judicial tendency in his dissenting
opinion in Lochner v. New York, 198 U.S. 45 pp. 74-75
(1904).

NOTES
to pages 85-86

29. See Louis L. Jaffe, "Suits Against Governments and Officers: Sovereign Immunity," 77 *Harvard Law Review* 1:1-39 (1963-1964).

30. See ibid., for a discussion of expansive versus restrictive interpretations of the act.

31. See the statement of one contemporary observer: "The solicitude of the oil companies for states' rights is hardly based on convictions derived from political theory, but rather from fears that federal ownership may result in the cancellation or modification of state leases favorable to their interests, their knowledge that they can successfully cope with state regulatory agencies, and uncertainty concerning their ability to control the federal agency." Professor Robert J. Harris quoted by Thomas R. Dye, *Politics in States and Communities* (Englewood Cliffs, New Jersey: Prentice-Hall, Inc., 1969), p. 55.

32. The United States v. California, 332 U.S. 19 (1947); holding that California owned no part of the tidelands.

33. For a good general discussion of prevailing attitudes, see Ernest R. Bartley, *The Tidelands Oil Controversy* (Austin: Univ. of Texas Press, 1953).

34. U.S. v. California, 381 U.S. 139 (1965).

35. Whatever Union's judgments as to the wisdom of paying without having to "fight" the claims of boat-owners and shoreline property owners for clean up costs, it is unlikely that the company would be willing to entertain the more far-reaching claims.

36. 43 U.S.C. Sec. 1333 (1964).

NOTES
to page 86

37. See particularly Calif. Water Code, Div. 7,
and Calif. Fish and Game Code, Div. 6, Chap. 2.

38. See Green v. General Petroleum, 205 *California
State Reports* (Cal.) 328; 270 *Pacific Reporter* (Pac.)
952 (1928).

39. Such a list by no means exhausts all possible
plaintiffs. Certainly the federal government could proceed against Union, under the Federal Refuse Act of 1899,
prohibiting dumping refuse into navigable waters. 33
U.S.C. Sec. 407 (1964). Under the Supreme Court's 1966
expansive definition of "refuse" to include virtually
all pollutants save sewage, Union should theoretically
be fair game for a Coast Guard complaint. See U.S. v.
Standard Oil Co., 384 U.S. 224 (1966); and contrast
with U.S. v. The Devalle, 45 *Federal Supplement* (F.
Supp.) 746 *Eastern District Louisiana* (E.D. La. 1942).
For an extensive discussion, focussing primarily on
pollution from tankers and other ships, see Joseph C.
Sweeney, "Oil Pollution of the Oceans," 37 *Fordham Law
Review*, 155-208 (1968), particularly 182-186. However,
due to: (a) Sweeney's discussion; (b) the relative im-
probability of such action given past and present Department of the Interior and Army Corps of Engineers
views on the desirability of Channel production; and (c)
limits of space, we shall leave the potential federal
governmental role as plaintiff unexplored. Could the
right state official profitably seek *mandamus* from the
right federal court requiring the right federal officer
to undertake prosecution? Probably not, see Mississippi
v. Johnson, 4 *Wallace* (Wall.) 475 (1866). For similar
reasons, we leave untouched potential plaintiffs such
as shipping and fishing interests of foreign nations,
and--except for brief mention *passim*--oil companies
miffed by stiffer federal regulations.

NOTES
to pages 86-87

40. See Note, "Airplane Noise, Property Rights and the Constitution," 65 *Columbia Law Review* 8:1428-1447 (1965), and particularly p. 1433 for occasional early injunctions against airplane noise.

41. John K. Galbraith, *The Affluent Society* (Boston: Houghton Mifflin Co., 1958).

42. Charles O. Gregory, "Trespass to Negligence to Absolute Liability," 37 *Virginia Law Review* 3:359-397 (1951).

43. Albert A. Ehrenzweig, "Negligence Without Fault," 54 *California Law Review* 4:1422-1477 (1966).

44. See Knodell, note 15 above, particularly p. 37ff. A fifth policy argument might be that the time has come for federal abandonment--or at least limitation--of sovereign immunity. See Jaffe, note 29 above. But the difficulties of giving that general position a local legal habitation and a justiciable name in the Santa Barbara oil slick cases, are too great to warrant discussion here, in light of the "discretionary" nature of the federal officers' "contributory actions". Similarly, there is little likely fruit for Santa Barbarans in the "proprietary-governmental" distinction. For its inapplicability to the U.S. government under the supremacy clause see New York v. United States, 326 U.S. 572 (1946); Graves v. New York ex. rel. O'Keefe 306 U.S. 466 (1939); Federal Land Bank v. Bismark, 314 U.S. 95 (1941); Carson v. Roane-Anderson Co., 342 U.S. 232 (1952). Consequently, we delete the federal government from the list of potential defendants. Our treatment is restricted to Union and her platform partners. (Hereafter for Union, read Union, Texaco, Mobil, Gulf.) Note, however, that in the Santa Barbara County and City "complaint for a mandatory injunction," dated April 4, 1969, and filed as No. 69-636-AAH in the U.S. District

Court for the Central District of California, Walter J.
Hickel and D.W. Solanas headed the list of defendants
named. Other defendants named were Union Oil, Pauley
Petroleum, Mobil Oil, Phillips Petroleum, Western Off-
shore Drilling and Exploration Company, Standard Oil,
Humble Oil, Peter Bawden Drilling Inc., Gulf Oil,
Texaco, and Atlantic-Richfield.

45. See Chief Justice White's opinion in Employ-
ers' Liability Cases, 207 U.S. 463, 502 (1908).

46. Prevailing assumptions seem to be that lengthy
suits will benefit Union, to the disadvantage of California plaintiffs. Perhaps so, if the plaintiffs have
difficulty meeting legal fees, and if quick favorable
results are possible. But perhaps from the standpoint
of ultimate success time may work the other way--allow-
ing similar crises to occur, the build-up of antipollution sentiment, and possibly, general overhauling
of tort doctrines.

47. The latter clause is not mere authorial vagueness. Rather, it anticipates another major arena of
genuine judicial confusion. See note 65 below.

48. Save possibly for numerous private litigants
in California small claims courts--whose annoyance value
might prove substantial--since failure of the defendant
to appear in court results in an automatic award to the
plaintiff, and in view of the prohibition against using
counsel. Since no opinions are written, it would be
hard for the defendant, having once won a similar case,
to prove *res judicata* (decided on its merits, and not
subject to litigation again by the same parties).

49. This was decided rather than facing the possibility of arguing jurisdiction all the way to the Supreme Court, losing, and having to start all over again.
See note 46 above as to that strategy's wisdom.

NOTES

to page 88

50. Assuming that the state initiates proceedings.
See U.S. v. California, 297 U.S. 175 (1936).

51. A nuisance must arise from a continuing situa-
tion--not from a "single occurrence." We stress "could"
find: It is not obvious to us that, without showing
clear danger of another blowout, a single act of pollution, despite its continuing nature, may be more appropriately remedied by damage suit. Possibly, then, the
time is not yet "ripe." Perhaps Santa Barbara must wait
for a second blowout. See above note 39, Sweeney, p.
181.

52. Interviews by A.E. Keir Nash in May 1969 and
June 1970. It seems, however, that his position on
Skiriotes v. Florida 313 U.S. 69. (1940) is not based
on solid ground. There the court allowed Florida to
enforce a state fishing law outside the three-mile
limit on the express grounds: (a) that Skiriotes was a
Florida citizen; and (b) that there was no federal law
or treaty in conflict with Florida's statute.

There are at least two difficulties involved in
attempting to draw a parallel here. (1) Though Union
is a California citizen, her "partners" on Platform A
are not. (2) Until a federal court determines that the
OCSL Act does not conflict with a California law, which
has been interpreted by a state court to require cessation of drilling on federal territory--an unlikelihood--
it would seem audacious to ignore a federal district
court ruling that it was defending its own jurisdiction
under 28 U.S.C. Sec. 283. See Note, "Immunity and
Public Authorities," 74 *Harvard Law Review* 4:714-725
(1960-1961).

But *contra*, Minier's action to vacate a temporary
federal restraining order--docketed as Civil Action No.

NOTES
to pages 88-89

69-712-S, heard June 20, 1969--relied heavily on U.S.
Supreme Court decisions that read Sec. 283 and its pre-
decessors narrowly, citing Hill v. Martin, 296 U.S. 393
(1935); Toucey v. N.Y. Life Ins. Co., 314 U.S. 118
(1941); Clothing Workers v. Richman Brothers, 348 U.S.
511 (1955); Stenfanelli v. Minard, 342 U.S. 117 (1951).
However, would the judicial policies of abstention from
interfering with state proceedings, enunciated in these
Supreme Court decisions, reach this situation? The most
hopeful case from the "Santa Barbara angle" is a 1960
decision by Judge John Minor Wisdom of the Fifth Circuit, T. Smith & Son, Inc. v. Williams, 275 *Federal
Reporter, Second Series* (F. 2d) 397, which denied injunctive relief sought by an employer from a state suit
brought by a longshoreman despite judicial conviction
that the federal Longshoreman's and Workman's Act
awarded exclusive jurisdiction in such cases to the
federal courts. According to Wisdom: "Further we think
that the broad language of Section 2283 and the policy
underlying it are against issuance of a federal injunction to stay state court proceedings even when the subject matter of an action, as federal courts see it,
rests exclusively in federal courts under a federal statute." Ibid., p. 407. Thus far the judges of the Ninth
Federal Circuit (which includes California) have not
agreed.

53. See William L. Prosser, "Private Action for
Public Nuisance," 52 *Virginia Law Review* 997 (1966).

54. Such a person would have to be carefully
chosen, however, as suffering a distinctly greater kind
of damage than the "average aggrieved Santa Barbaran."
A renter of pleasure boats?

55. See note 65 below.

56. See Comment, "Torts Along the Water's Edge,"
1968 *Illinois Law Forum* 1:95-103.

NOTES
to pages 89-90

57.   David v. City of Jacksonville Beach, 251 F.
Supp. 327 (M.D. Fla. 1965).

58.   Pope and Talbot, Inc., v. Hawn, 346 U.S. 406,
74 *Supreme Court Reporter* (S. Ct.) (1953).   Thus a potential defense in civil jurisdiction against claimant
boatowners--failure to move boats from the Santa Barbara Harbor during the days before the arrival of the
slick--would not help in admiralty.

59.   King v. Testerman, 214 F. Supp. 335 (E.D.
Tenn. 1963). But *contra*, Gaderson v. Texas Contracting
Company, 3F.2d 140 (Fifth Cir. 1924) whose influence
has perhaps diminished over time.

60.   Movible Offshore Company v. Ousley, 346 F.2d
870 (Fifth Cir. 1965).

61.   Rodrigue v. Aetna Casualty and Surety Company,
395 U.S. 352 (1969).

62.   Luthringer v. Moore, 181 *Pacific Reporter,
Second Series* (P. 2d) 89 (1947).   Fumigator of one
building held liable for damage to adjacent building.
For a related discussion favoring general state legislative codification of imposing on the developer liability for water run-offs from the development of land,
see Note, "California's Surface Waters," 39 *Southern
California Law Review* 1:128-136 (1966).

63.   205 Cal. 328; 270 Pac. 952 (1928).

64.   See note 70 below.

65.   Palsgraf v. Long Island Railroad, 248 *New
York State Reports* (N.Y.) 339, 162 *New England Reporter*
(N.E.) 99 (1928).

NOTES
to page 91

66. 43 Stat. 604 Chap. 316, Sec. 8.

67. This issue appears still in the process of
settlement in lower federal courts. Sirmons v. Baxter
Drilling Inc., 239 F.2d 348 (1965) (an anchored platform not a vessel); Producers Drilling Co. v. Gray, 361
F.2d 432 (1966) (a drilling barge, anchored, is a vessel).

68. The New York Harbor Act of 1888, 33 U.S.C.
Secs. 441-451b (1964), forbade discharge of refuse into
New York, Baltimore and Hampton Roads Harbors. The
Refuse Act of 1899, 33 U.S.C. Secs. 407-409, 411, 419
(1964), prohibited dumping refuse into navigable waters.
Both of these acts applied to vessels and shore facilities. The Federal Water Pollution Control Act, 33
U.S.C. Sec. 46ff. (1964), required the establishment of
water quality standards for all interstate waters, and
of plans for implementing and enforcing such standards.
Additionally, the act subjected to abatement pollution
of interstate or navigable waters that endangers the
public health or welfare. The Oil Pollution Act of 1961,
33 U.S.C. Secs. 1001-1015 (1964), implemented the International Convention for the Prevention of Pollution of
the Seas by Oil, 1954, which restricted the discharge
of oily wastes to certain international waters. The
Convention on the Territorial Sea and the Contiguous
Zone, Article 24, Geneva 1958 (entered into force September 10, 1964) provided that a coastal state may, in
the contiguous zone not more than 12 miles from shore,
exercise the control necessary to prevent infringement
of customs, fiscal, immigration, or sanitary regulations within its territory. See Chapter VIII below for
discussion of legislation enacted or proposed since the
date of the Santa Barbara spill.

69. 294 F. 597 (Eighth Cir. 1923).

NOTES
to page 92

70. The leasing and operating regulations for the
Outer Continental Shelf in effect on January 28, 1969
would appear to support this conclusion. See particularly, 30 C.F.R. Sec. 250.42: *Control of Wells*. (a)
The lessee shall take all reasonable precautions for
keeping all wells under control at all times....(b) The
lessee shall take all reasonable precautions to prevent
any well from blowing open and shall take immediate
steps and exercise due diligence to bring under control
any such well. *Outer Continental Shelf Order No. 2*
signed by D.W. Solanas. *Outer Continental Shelf Order
No. 7* (also issued by Solanas on the same date--March
31, 1965) might appear to proceed further by bringing
lessee's attention to 30 C.F.R. Sec. 25D. 42 *Pollution*.
"The lessee shall not pollute the waters of the high
seas or damage the aquatic life of the sea...."

71. There are at least seven points where negligence might be alleged: (1) drilling in the Channel
at all; (2) Union's request for a variance; (3) the
blowout itself; (4) permitting the oil to get within
three miles and thus reaching California territory; (5)
reaching shallow waters where crustaceans, plankton, and
other organisms are affected; (6) reaching the harbor
and thus touching property in moored boats and city-
owned installations; (7) touching real property on land.

One telling argument might be, if it could be
proved, that both earlier difficulties encountered by
Union in the Dos Cuadras lease and an industry rule-of-
thumb about safe drilling practices should have led
Union to set more intermediate casing lengths. That
rule-of-thumb states: (a) that the pressure of oil-
bearing strata generally increases at about 16 lb. per
sq. in. for every vertical foot drilled; and (b) that
the capacity of rock strata to resist fracturing by high-
pressure oil is about twice that. Thus, one might expect at the bottom of Well A 21's casing some 250 feet

NOTES
to page 92

below the ocean floor--a resistance capacity of perhaps
500 lb. per sq. in., and a pressure of 4 or 5 times
that at the bottom of the well. Any attempt to shut
off the well once oil began to flow might, hence, be
likely to fracture the strata near the bottom of the
casing. Hence one might anticipate leakage even in
the absence of pre-existing faults.

72. See Robins Dry Dock v. Flint, 275 U.S. 303
(1927) holding that damage to a charter boat during dry-
dock storage was recoverable, but that consequent loss
of income was not. See also, The Menominee, 125 F. 530
(E.D. N.Y. 1903), denying damages for lost fishing time
and thus lost fish, due to damage to fishing boat. Contrast, Carbone v. Ulrich, 209 F.2d 178 (9th Cir. 1953)
which allowed recovery for lost prospective catch of
fish; and Small v. United States, 333 F. 2d 702 (3rd
Cir. 1964), for a "winning" argument that cancellations
at a beachfront resort were consequential to property
damage and thus recoverable. For a more detailed discussion, see above note 39, Sweeney, pp. 174-175. We
would query further, does Palsgraf need "updating?" A
less severe problem for "non-littoral" parties than
foreseeability would appear to be reckoning of damages.
See Territory Illuminating Oil Co. v. Townley, 81 F. 2d
159 (10th Cir. 1956). If the cause can be established,
the damages do not evidently have to be calculated
exactly. Presumably, proof that lost profits could not
have been less than "X" dollars, would allow recovery
up to that amount.

152

NOTES
to pages 93-95

1. See the provocative but perhaps overdrawn analysis of Harvey Molotch, "Santa Barbara: Oil in the Velvet Playground," *Ramparts* 8(5):43-51 (1969).

2. That is to say, regardless of whether the "Lowian" or "Wilsonian" critiques appear more applicable.

3. See Lowi, Chapter I, note 34. Lowi's emphasis is on partial decisions in the sense of incomplete decisions--on the "non-cumulativeness" of governing experience when there is bargaining about each decision without a guiding set of rules and cumulative, coherent precedents.

4. See S. 2516, 91st Cong., 1st Sess., introduced June 30, 1969, by Senator Dirksen, for Senator George Murphy.

5. See note 3 above. See especially Lowi's pp. 55-67, where he argues that since 1937, with the Supreme Court's "switch in time that saved nine," the liberal-conservative dialogue has been an empty debate due to the effective consensus which then emerged on the virtues of what he calls "Interest-Group Liberalism." Before 1937, he argues there was a genuine debate between liberals and conservatives characterized by "unanimity on the underlying criteria" for judging the two positions--"attitude toward government...and toward social change, or 'planning.'" Since then, he argues, "The old dialogue has passed into the graveyard of consensus. Yet it persists....despite its irrelevance....(which) has not been without many evil effects. This persistence has blinded the nation to the emergence of a new and ersatz political philosophy. The coexistence of a purely ritualistic public dialogue with an ersatz, unappreciated, uncriticized, but quite real new public philosophy has produced most of the political pathologies of the 1960's. The empty rhetoric of liberalism-conservatism has meant the decline of meaningful

NOTES
to pages 95-98

adversary proceedings in favor of administrative, technical, and logrolling considerations...." pp. 56-57.

On this showing he who constitutes the largest
group dominates the logrolling, and he who dominates
the logrolling dominates the decision.

6. Santa Barbara *News-Press*, October 20, 1969,
Sec. A, p. 9.

7. See *Hearings Before the Subcommittee on Minerals, Materials, and Fuels*, March 13-14, 1970 (Washington, D.C., 1970), p. 15.

8. Ibid., p. 84. Testimony of Norman Sanders,
President of the Western Citizens for Environmental
Defense, Inc.

9. See Chapter VI, note 33, and related text.

10. L.A. *Times* editorial reprinted in the Santa
Barbara *News-Press*, June 17, 1970, Sec. H, p. 12. For
denial of drilling applications, see Los Angeles *Times*,
September 21, 1971, Sec. I, pp. 1 and 18.

11. Ibid., p. 12.

12. "Statement of Walter J. Hickel, Secretary of
the Interior, on S. 4017, before the Subcommittee on
Minerals, Materials, and Fuels of the Committee on Interior and Insular Affairs of the United States Senate."
Delivered orally on July 23, 1970. Obtained in unpub-
lished form from the Interior and Insular Affairs Committee Office by A.E. Keir Nash on July 23, 1970, p. 6.

13. Ibid., p. 6.

14. Santa Barbara *News-Press*, June 21, 1970, Sec.
A, p. 8.

NOTES
to pages 98-100

15. Daniel Moynihan, "Policy vs. Program in the
'70's," *The Public Interest*, No. 20:90-100 (Summer 1970).
Note that Moynihan prefers the term "policy" to the
politically less acceptable term "planning." Is this
an illustration of Lowi's point that a ritualistic public pseudo-dialogue continues? See note 5, above. Un-
doubtedly, Mr. Moynihan would understand the point.

16. *Toward Balanced Growth: Quantity with Quality:
Report of the National Goals Staff* (Washington, D.C.:
1970). For further analysis of this report, see James
M. Naughton, "Panel Finds Need to Inspire Debate on
Nation's Goals," New York *Times*, July 19, 1970, p. 1.

17. Samuel Beer, "Great Britain: From Governing
Elite to Organized Mass Parties," in Sigmund Neumann,
ed., *Modern Political Parties* (Chicago: University of
Chicago Press, 1956) p. 9, quoting Keith Feiling.

18. P.L. 91-224 Title I, Sec. 11(b).

19. S. 3471, 91st Cong. 2d Sess. (1970), introduced by Minority Leader Senator Hugh Scott of Pennsylvania.

20. Ibid., particularly Sec. 10(4)(f).

21. S. 3687, 91st Cong. 2d Sess., introduced
April 7, 1970.

22. Ibid., Sec. 7(d).

23. Ibid., Sec. 10(b)(2).

24. S. 3181, 91st Cong., 1st Sess., introduced
November 25, 1969.

25. See Chapter VI, note 23.

26. See note 5 above.

27. Ibid., particularly pp. 147-153.

28. Ibid., particularly, pp. 125-156.

29. So, at least, seemed to think a highly placed aide in the office of one of the senators most involved in the controversy over offshore oil. Interview by A.E. Keir Nash, July 23, 1970, New Senate Office Building, Washington, D.C.

30. Santa Barbara *News-Press*, December 17, 1969, Sec. A, p. 1.

31. Santa Barbara *News-Press*, April 23, 1970, Sec. A, p. 1.

32. Santa Barbara *News-Press*, January 11, 1972, Sec. A, p. 1.

33. See Chapter V, note 3. For a contemporary development of such an assumption, see Grant McConnell, *Private Power and American Democracy* (New York: Alfred A. Knopf, 1951), particularly p. 190.

34. 295 U.S. 495 (1935).

35. 298 U.S. 238 (1936).

36. United States v. Causby, 328 U.S. 256 (1947).

37. Justice Owen Roberts dissenting in Smith v. Allwright, 321 U.S. 649, p. 666 (1944).

38. Samuel Huntington, *Political Order in Changing Societies* (New Haven: Yale University Press, 1968), p. 99.

NOTES
to pages 106-109

39. See Louis Hartz, *The Liberal Tradition in America* (Boston: Little, Brown, 1955), where Hartz gives in advance a good reason for the "fixation" on an empty liberal-conservative debate in the contemporary political arena discussed by Lowi. For Hartz, American society was born to the heritage of John Locke, and without competing political philosophic traditions, has remained "stuck" with him.

40. "The Role of Government in a Free Society," in Marshall I. Goldman, ed., *Controlling Pollution* (Englewood Cliffs: Prentice-Hall, Inc., 1967), pp. 73-81, p. 74.

41. Ibid.

42. Ibid.

43. (New York: Charles Scribner's Sons, 1968), pp. 144-145.

44. "The Next Industrial Revolution," *Science* 167(3926):1673(March 27, 1970).

45. Loc. cit.

46. Loc. cit.

47. See Chapter IV, note 10 above.

48. Karl William Kapp, *Social Costs of Business Enterprise* (New York: Asia Publishing House, 1963).

49. That is to say, the problem of "solving pollution" seems to require generating the pressure and the demand to solve a prior, existing problem: poverty. For an insightful discussion of "backward linkage" and "forward linkage" in development economics, see Albert

NOTES
to pages 110-112

O. Hirschman, *The Strategy of Economic Development* (New Haven: Yale University Press, 1958).

50.   See the President's two messages of July 9, 1970, to Congress, 91st Cong., 2d Sess., Documents 91-364 and 91-365, respectively.   In the absence of contravening action by Congress, Reorganization Plan No. 3 of 1970 establishes the Environmental Protection Agency, which would be the central administrative arm of the federal government for dealing with environmental problems.   At its head is an agency administrator who would have command over functions now exercised by the Federal Water Quality Administration, and by sundry components of the Department of Interior, HEW, Agriculture, and the Atomic Energy Commission having to do with the environment.   Reorganization Plan No. 4 of 1970 sets up a National Oceanic and Atmospheric Administration in the Department of Commerce.   Both are aimed at centralizing administrative actions relating to the environment.   At least, Plan No. 3 goes a considerable distance in taking away from departments areas of power where they may have tended to "collude" rather than to "regulate."

51.   For such an interpretation see Los Angeles *Times*, September 21, 1971, Sec. 1, pp. 1 and 18.

52.   Santa Barbara *News-Press*, December 30, 1971, Sec. B, p. 1.

53.   Such is the thrust of Senator Cranston's recent amending of S. 1219.

54.   Comment, "Pre-emption," 10 *Arizona Law Review* 1:97-106 (1968), especially pp. 104-106.

55.   381 U.S. 479 (1965), striking down Connecticut's 1879 law forbidding the dissemination of birth control devices.